# A Cosmist Manifesto

## Practical Philosophy for the Posthuman Age

Ben Goertzel

**humanity⁺**

# PRESS

`humanityplus.org/press`

# Dedication:

# to Valentin Turchin

This Manifesto is dedicated to **Valentin Turchin** (1931 - 2010), a great Soviet-American scientist and futurist visionary who died the year it was completed.

I'm very sorry Val never got to read this Manifesto, as I'm sure he would have enjoyed it. He would have agreed with most of it, and had insightful and entertaining arguments to make about the rest. While I never explicitly discussed "Cosmism" with him, I have rarely met anyone more Cosmist in their attitudes, through and through.

As a cybernetician and computer scientist, Val's contributions were numerous, including the Refal programming language, the theory of metasystem transitions, and the notion of supercompilation. He was a pioneer of Artificial Intelligence and one of the visionaries at the basis of the Global Brain idea.

And his book The Phenomenon of Science, written in the 1960s, is one of the most elegant statements of Cosmist scientific philosophy ever written.

I was privileged to know Val in the late 1990s and early 2000s, when we both lived in North Jersey, in the context of collaborating with him on the commercialization of his supercompilation technology. Our discussions on supercompilation, immortality, AI, the philosophy of mathematics and other topics were among the most memorable I've had with anyone.

The death of great minds like Val is one of the absurd horrors that Cosmist philosophy hopes to abolish via scientific and technological advance.

The Phenomenon of Science closes with the following words:

"We have constructed a beautiful and majestic edifice of science. Its fine-laced linguistic constructions soar high into the sky. But direct your gaze to the space between the pillars, arches, and floors, beyond them, off into the void. Look more carefully, and there in the distance, in the black depth, you will see someone's green eyes staring. It is the Secret, looking at you."

# Contents

## Preface

The basic theme of this book –  a practical philosophy encompassing humankind's quest to comprehend, experience and perhaps ultimately fuse with more and more of the universe -- took hold in my mind in my early youth, sometime between the ages of 6 and 9, when I was reading both a host of classic science fiction novels, and bits and pieces of Asian history, such as Will Durant's *History of Civilization* and one of my mother's grad school texts discussing the life of Buddha. It struck me as both obvious and fascinating that the SF writers and the Eastern sages were pushing in the same direction: beyond boring, annoying old everyday human reality, toward different kinds of experiences, knowledge and understanding. Beyond the routine and out into the Cosmos!

All those SF stories about AI, immortality, alien minds, nanotech, virtual worlds and all that –  they weren't just about gee-whiz technology. They were about the nature of the world. They were about how there may be a hell of a big universe out there –  most of which is probably well beyond the grasp of our tiny little human selves, minds and societies. And they were about the possibility of us coming to grips with some of these other, richer aspects of the universe –  and maybe becoming something better than "us" in the process.

And all the Buddhism stuff I read (and, 2 or 3 years later, works by Ouspensky and others  in a similar vein) wasn't just about meditation or other strange Eastern religious practices.  It was about the nature of the world and the mind –  the way the  mind got built up by observing the world; and the world got built up by the observing mind.  It was about how the things most people thought were very important, maybe weren't so critical after all.  About the universe being different, and more (and in some senses less: e.g. less rigid and independent), than we commonly think it is.

Fast forward a few decades, and in the late 1990s, in my mid-30s, I became involved with the futurist community –  transhumanists, Singularitarians, AI visionaries, etc.  I had spent many of the intervening years working on artificial general intelligence and other out-there futurist ideas and technologies, but mostly on a solo basis.  It wasn't till around the turn of the century that I discovered a community of other wild-eyed visionaries who shared both my excitement about advanced technology, and my faith that this tech was about more than just gadgetry –  that it provided a window into some fundamental things about mind, reality and existence.

By now there are a lot of good nonfiction books about advanced technologies and their likely future development trajectories, and their social and other implications. But as I become more ensconced in the futurist community, I became increasingly frustrated at the lack of any text presenting likely-fairly-near-future technologies and their psychological, social and philosophical implications from a really *fundamental* perspective.

I.e.: what would the advent of such technologies *really mean*? What would it teach us about the nature of ourselves and the world? What kind of world-view might suffice to comprehend a world dominated by such technologies and their successors – or to guide us through the transition to such a world? What network of ideas might form the seed about which some sort of active community of futurists might crystallize, pushing themselves and one another to push harder and harder toward a radically transhuman future?

I had written a bit about these themes in my 2006 philosophy tome *The Hidden Pattern*, and 10 years before that in a work called *The Unification of Science and Spirit*, which is on my website but was never published because I never became happy with it (and never got around to suitably revising it). Long before that, at age 19 or so, I had written a work variously titled *Transnihilistic Visions* and *?*, with related core ideas. But none of those works treated the intersection of advanced technology and fundamental philosophy as directly, crisply and *pragmatically* as I thought the topic deserved.

So, in mid-2009 I set out to write a few pages of coherent notes on the topic -- and wound up writing this little book instead.

The book began as a series of blog entries -- initially written in a few 18-hour days at the computer, and then edited and expanded in bits and pieces over the succeeding months. Depending on when you're reading this, the blog may still be online at cosmistmanifesto.blogspot.com. The text there is a rough and partial draft of the text here, but the comments made by readers there may be of some additional interest. Feel free to log on there and post additional comments!

The term "Cosmism" was chosen basically for lack of a better one.

"Transhumanism" and "Posthumanism" place the focus on humanity which didn't feel right. And the same for "H+" : I am a Board member (and former Chairman) of the organization Humanity+ (H+), and am very supportive of the H+ movement ... and there is no particular contradiction between H+ and Cosmism ... but the general focus of the H+ meme is a bit different than what I want to emphasize here. It wouldn't seem right to call this book " An H+ Manifesto" because so many of the ideas here go so far beyond the less-radical human-enhancement themes that for many are the crux of H+.

"Singularitarianism" commits to a particular class of future pathways, which are interesting but not critical to the main points I want to make. "Futurism" is way too broad.

The previous users of the term "Cosmism" held views quite sympathetic to my own, so classifying my own perspective as an early 21st century species of Cosmism seems perfectly appropriate.

And thus: *A Cosmist Manifesto.*

## Who Is This Book For?

The book is for everyone who likes thinking and understanding. For everyone who wants to understand their world, their mind, or their future.

But it will go down most easily for the reader who's already absorbed a bit of technofuturism -- perhaps from reading modern SF writers like Vinge or Stephenson, Stross or Broderick; or perhaps from the nonfiction works of Kurzweil, Broderick (again) or Drexler or other futurist pundits.

There is a lot to say about Cosmism, but in writing this Manifesto I've aimed for compactness over completeness -- not only because I have a lot of other things to do than write manifestos, but also because I want to make sure the focus is on the essentials.

As a result of its compactness, this brief work is probably not terribly "novice-friendly" -- if you've not plunged into the early 21st century techno-futurist literature at all before, you may find it perplexing and opaque, and you may want to do some other reading first and come back to this a little later. Or not -- sometimes it's best to just plunge in!

## Many Debts, Few References

In fact, the vast majority of ideas presented here are things I've written down before in one book, article, essay or another, over the years, often in much more depth than is done here. But most of those prior writings have been aimed at an academic audience; and I've sometimes felt that in those writings some of the core ideas have been expressed with inadequate clarity due to the various connections and complications elaborated therein. Sometimes there is power in simplicity.

There are many, many details pertaining to all the points raised here, and exploring them is critical -- but it's also critical to be clear on the fundamentals and not to get lost in the particulars.

In this spirit, you'll notice an absence of references and citations in this text. Having suffered through the writing of around 100 scientific journal papers and the writing and editing of 12 academic books, I know how to write in a fully-referenced academic style all too well -- and this is intentionally not that kind of work. However, by avoiding academic-style referencing, I'm certainly not representing that every idea presented here is original. Some are original; many are not! Sometimes I mention another historical or contemporary thinker by name, when it seems particularly appropriate -- but these mentions are not particularly systematic and don't necessarily reflect the biggest influences or sources of the ideas given here.

## The Author's Hope

My hope is that you'll find the practical philosophy I articulate here not only interesting but also compelling – and maybe even useful. Cosmism isn't just about cool ideas that are fun to think, talk and write about. It's about actively trying to understand more, actively trying to grow and improve and collectively create a better cosmos, and all that good stuff...

As will become clear to you if you read the rest of this Manifesto, one aspect of Cosmism is, that, roughly speaking: the more sentient beings adopt Cosmist values, the better will Cosmist values be served.

Of course, I don't expect anyone to fully agree with everything I say here — I myself, in a decade or a year or maybe even a month, may not agree with all of it!

However, if you agree with a substantial percentage of Cosmism as I articulate it here — and more importantly, if you agree with the *spirit* in which these thoughts are offered — then you are a Cosmist in the sense in which I mean the term.

# A Few Acknowledgements

I won't try to acknowledge everyone whose intellectual input over the years helped form the perspective presented here. There have been literally hundreds – via writings, and online or F2F conversations. Some of you know who you are. Thank you all, so very very much!

I must mention my family who has listened to me spout these ideas way too many times: my late grandfather Leo Zwell, my father Ted Goertzel, my mother Carol Goertzel; my kids Zarathustra, Zebulon and Scheherazade; my first wife Gwendalin Qi Aranya and my second wife Izabela Lyon Freire.

Next I must thank two people whose encouragement led directly to the production of this book (and who are both quoted herein): Giulio Prisco and Samantha Atkins.

And the futurist organizations I've been involved with: the Immortality Institute, the Singularity Institute for AI, and Humanity+ (formerly the World Transhumanist Association). Though I must stress that the views presented in this book are mine and overlap to varying degrees with the official views of these organizations.

Finally, thanks are due to everyone who commented on the early blog versions of the book (some in the blog comment areas, some via email), thus providing useful feedback that shaped the present version.

And I am grateful to Natasha Vita-More for creating and contributing the wonderful cover art.

Any mistakes or foolish statements in the book are entirely the fault of the Cosmos.

# What Is Cosmism?

By **Cosmism** I mean: **a practical philosophy focused on enthusiastically and thoroughly exploring, understanding and enjoying the cosmos, in its inner, outer and social aspects**

Cosmism advocates

- pursuing joy, growth and freedom for oneself and all beings
- ongoingly, actively seeking to better understand the universe in its multiple aspects, from a variety of perspectives
- taking nothing as axiomatic and accepting all ideas, beliefs and habits as open to revision based on thought, dialogue and experience

The word "Cosmism" has been used by others in the past in various ways, all of them related to and fairly harmonious with the sense in which I mean it here ... I will review these briefly below; but in this Manifesto I'm largely ignoring the particulars of these prior uses.

My goal in this Manifesto is to clearly and simply articulate *my own* take on Cosmism — that is: the particular flavor of Cosmism that I find most sympathetic.

I've said Cosmism is a practical philosophy. What I mean by a "practical philosophy" is, in essence, a **world-view and value-system** — but one that, in addition to containing abstract understanding, provides concrete guidance to the issues we face in our lives.

Like any other world-view and/or value system, Cosmism is not something that can be scientifically or mathematically proven to be "correct"; it is something that an individual or group may adopt, or not. Obviously I think Cosmism worthy of adoption, or I wouldn't be writing a Manifesto about it.

Not only do I think Cosmism is a Good Thing in a general sense — I think it will become increasingly relevant in the next years, decades and centuries as technology advances, as the "human world" we take for granted is replaced with a succession of radically different realities.

The currently standard world-views and value-systems will, I suggest, not only fail to survive this transition, but — worse yet — fail terribly as guides as we pass through it. Cosmism is far better suited to guide us as these changes unfold.

It appears likely to many knowledgeable people — including me — that advanced science and technology will soon allow our minds to expand far beyond the limitations of the human brain architecture that has historically supported them.

I won't take up space repeating the evidence for this assertion here: Kurzweil's *The Singularity Is Near* and Broderick's *The Spike* are good places to start if somehow you've found this Manifesto without first being familiar with the canon of modern futurism.

Cosmism would be an interesting and relevant philosophy even without this dawning technological Singularity/ Transcension.

However, these probable impending events make Cosmism more appealing — for the reason that the alternative philosophies more prevalent among the human race at the current time, are deeply incompatible with the changes that are coming.

Cosmism provides a world-view and value-system that makes sense in the human world now, and will continue to make sense as the practical world advances, even as some of us leave our human bodies and brains behind and explore new ways of existing and interacting.

Quite possibly once we become advanced enough, Cosmism will appear to us roughly equally silly as all other "legacy human" philosophies. If so, then I doubt I will be shedding any tears for it at that stage! But I will be happy that it proved adequate to help nurse us through the transition to our next phase of being. (Although, even if some continuous evolution of mine is around at that time, it's unclear whether it will still identify itself as being the same "self" or mind as Ben Goertzel circa 2010!)

If your main interest is in Cosmist views of future technologies, you may wish to skip toward the middle of this Manifesto where they're explicitly treated. But eventually you'll probably want to look back at the earlier parts, which outline the philosophical foundation on which the later more tech-focused discussions are based.

Every one of the radical future technologies dawning has profound philosophical implications, going beyond what is explored in SF movies and all but the most profound SF books. Understanding these technologies and what they will do and what they will mean requires taking a deep look at the nature of the mind and the universe.

Just as the Internet is about people as much as it is about wires and bits and protocols, the new technologies dawning are about mind and reality as much as they are about AI algorithms, engineered gene sequences and nanodevices.

Understanding **artificial intelligence** — and the sense in which it may be sentient — requires us to look deeply into the nature of mind and awareness.

Understanding **brain-computer interfacing** requires us to deeply understand the mind and the self, and their relation to other minds and to physical reality.

Understanding the emerging **global brain** requires us to understand the nature of mind and society in a way that goes beyond the models we conventionally use, which are based on current biological brains and societies that will soon be dramatically augmented or transcended.

Understanding **immortality** and the issues associated with it requires an understanding of self and identity — of what is a "self" that it might be immortal.

Exploring the various possible means to immortality — including **uploading** and other forms of **cyberimmortality** — requires an understanding of the relations between minds and bodies.

Understanding what advanced **unified physics** might mean requires deeply understanding the nature of physical reality, including subtle issues like the relation between reality and simulation.

Understanding what **virtual realities** or inexpensive **molecular assembler**s would mean for human or more advanced forms of life and mind, again requires a profound understanding of the interrelation of mind, reality and society.

Understanding what sorts of **alien minds** we might discover — elsewhere in the physical universe, in other "dimensions" or potentially right here on Earth — requires a deep understanding of mind, reality and their relationship.

Thinking about these possibilities from a purely technological perspective is inadequate and may perhaps be dangerously misleading. These possibilities must be considered very deeply from a perspective of pragmatic philosophy, if one is to have any real hope of understanding and approaching them in a useful way. That is one of the key things that Cosmism, as I interpret and pursue it here, attempts to do.

So, in the first N sections of this Manifesto I'm going to delve fairly deep into what will seem like abstract philosophical considerations. But it all will get pulled back into the practical by the end.

## Ten Cosmist Convictions

*(Mostly Suggested by Giulio Prisco)*

Giulio Prisco, on the mailing list of a group called the "Order of Cosmic Engineers", posted a wonderful "mini-manifesto" listing principles of the OCE. I have edited and extended his list slightly, without altering its spirit, to obtain the following, which may serve as a reasonable preface to this Manifesto:

1)  Humans will merge with technology, to a rapidly increasing extent. This is a new phase of the evolution of our species, just picking up speed about now. The divide between natural and artificial will blur, then disappear. Some of us will continue to be humans, but with a radically expanded and always growing range of available options, and radically increased diversity and complexity. Others will grow into new forms of intelligence far beyond the human domain.

2) We will develop sentient AI and mind uploading technology. Mind uploading technology will permit an indefinite lifespan to those who choose to leave biology behind and upload. Some uploaded humans will choose to merge with each other and with AIs. This will require reformulations of current notions of self, but we will be able to cope.

3) We will spread to the stars and roam the universe. We will meet and merge with other species out there. We may roam to other dimensions of existence as well, beyond the ones of which we're currently aware.

4) We will develop interoperable synthetic realities (virtual worlds) able to support sentience. Some uploads will choose to live in virtual worlds. The divide between physical and synthetic realities will blur, then disappear.

5) We will develop spacetime engineering and scientific "future magic" much beyond our current understanding and imagination.

6) Spacetime engineering and future magic will permit achieving, by scientific means, most of the promises of religions — and many amazing things that no human religion ever dreamed. Eventually

we will be able to resurrect the dead by "copying them to the future".

7) Intelligent life will become the main factor in the evolution of the cosmos, and steer it toward an intended path.

8) Radical technological advances will reduce material scarcity drastically, so that abundances of wealth, growth and experience will be available to all minds who so desire. New systems of self-regulation will emerge to mitigate the possibility of mind-creation running amok and exhausting the ample resources of the cosmos.

9) New ethical systems will emerge, based on principles including the spread of joy, growth and freedom through the universe, as well as new principles we cannot yet imagine

10) All these changes will fundamentally improve the subjective and social experience of humans and our creations and successors, leading to states of individual and shared awareness possessing depth, breadth and wonder far beyond that accessible to "legacy humans"

## P.S.

Giulio Prisco, who formulated the first draft of the above list, made the following comment on the use of the word "will" in these principles:

*" ... 'will' is not used in the sense of inevitability, but in the sense of intention: we want to do this, we are confident that we can do it, and we will do our f\*\*king best to do it."*

## A Brief History of "Cosmism"

The term "cosmism" seems to have originated with the Russian Cosmists, in the mid-1800's.

The most famous Russian Cosmist was Konstantin Tsiolkovky, who according to Wikipedia

*believed that colonizing space would lead to the perfection of the human race, with immortality and a carefree existence. He also developed ideas of the "animated atom" (panpsychism), and "radiant mankind".*

All this is generally conceptually harmonious with my use of the term here, though not precisely identical.

My friend and colleague Hugo de Garis has used the term Cosmist to refer to (again quoting the mighty Wikipedia)

*a moral philosophy that favours building or growing strong artificial intelligence and ultimately leaving the planet Earth.... Cosmists will foresee the massive, truly astronomical potential of substrate-independent cognition, and will therefore advocate unlimited growth in the designated fields, in the hopes that "super intelligent" machines might one day colonise the universe. It is this "cosmic" view of history, in which the fate of one single species, on one single planet, is seen as insignificant next to the fate of the known universe, that gives the Cosmists their name.*

Again this is generally harmonious with my use of the term here, though not precisely identical.

Different people have used the term Cosmism with different shades of meaning, but we're all pushing in the same general direction!

## Cosmism versus Other Futurist Memes

Cosmism, futurism, Singularitarianism, H+, Extropianism, transhumanism, accelerating change … there are a lot of futuristic memes and dreams floating around these days, some with conferences or organizations associated with them. What's the difference really?

To me that is mainly an issue for future sociologists with too much time on their hands. The important thing is the emerging network of ideas and realities that *all* these different memes collectively get at, in their own different ways. A petty sectarianism has plagued many promising social and intellectual movements in the past, and we shouldn't let it pollute the 21st century futurist community.

Then why use "Cosmism" instead of some other, currently trendier term? It's purely a matter of emphasis.

I am heavily involved with the H+ and Singularitarian memes, being a Board member (and former Chairman) of the organization Humanity+, and former Director of Research of the Singularity Institute for AI. I support these networks of ideas and people very much. But each has a different emphasis.

The Singularity is a particular hypothetical event. This is not a Singularitarian Manifesto because most of the points made here are relevant whether or not there is a Singularity; for instance in the case where technology triggers amazing changes in mind and reality, but more gradually.

H+ (which I consider as basically synonymous with transhumanism, though some others may disagree) places more focus on nearer-term, less radical advancements like smart drugs and artificial organs and limbs. Cosmism embraces these things but it doesn't focus on them, any more than the World Explorers Club focuses on the genuinely wonderful things you can see walking around in your garden. H+ does have a place for the wilder possibilities that Cosmism highlights, but it doesn't position them center stage.

And H+ is narrower than Cosmism in the sense that it doesn' t deal with notions like panpsychism, or the possibility of a universal mind, or the inter-construction of mind and reality. H+ doesn' t reject these things out of hand, but they' re not central to its enterprise – whereas, they are part and parcel of the understanding that Cosmism seeks to build.

My feeling is that H+ is too broad to serve as the basis for a coherent world-view – it is more of an umbrella notion, with the power to encompass a host of interrelated world-views, Cosmism being one of them. Whereas Singularitarianism focuses on a single event and its likely precursors and aftermath, and is hence too narrow to serve as the basis for a coherent world-view.

Cosmism, as I construe and present it, does not aim to contradict these other futurist memes, but rather to complement them with its own flavor of understanding. I believe that Cosmism has the right level of specificity to serve as the basis of a coherent, practical philosophy – without sinking to the level of dictating specific near-term positions like a political platform, and also without leaving major issues unresolved and ambiguous.

In other words, my thought is that a fully developed Cosmism would

- answer the Big Questions just about as well as is possible within the scope of human intelligence

- leave it to individual Cosmists to resolve the multitude of small questions after their own fashion, utilizing the Cosmist memeplex as a conceptual toolkit

Of course this book does not constitute a fully developed Cosmism, but I hope that it's a nontrivial step.

## Cosmism as Old-Style Philosophy

While this Manifesto focuses on some highly modern (post-modern? post-postmodern?) ideas, in some ways it's quite old-fashioned.

It reflects an approach to philosophy that was more common before 1950, before philosophy become so academic and formal.

My methodology here is much like that of Nietzsche, or Schopenhauer, or Plato, or Lao Tzu — I'm presenting my overall understanding of life and the world, with a view toward practical guidance as well as conceptual understanding. Greatly inspired by the prior ideas of others, but also with a strong personal slant.

Robert Pirsig and Paul Feyerabend are two fairly recent philosophers who inspired me with their direct, "old fashioned" approach to philosophy — writing simply rather than technically, and giving their holistic understanding rather than focusing on painstaking disection of micro-issues. I think we need more philosophy of this sort, which is one of the (many) reasons I wrote this book.

## Cosmism versus Academic Philosophy

Most of the "philosophy" done by professional philosophers today involves complex, abstract and refined approaches to deeply understanding particular aspects of the world, in a highly precise and intellectual way.

Cosmism does have this aspect — but it's not the aspect I'm going to emphasize here. I will touch on abstract topics as necessary (especially in the first third of the text, in which I strive to articulate the deep conceptual foundations of Cosmism), but by and large I'll move past them fairly quickly to get on to more concrete stuff.

I'm going to mostly focus on Cosmism as a practical philosophy for living ... both now, and (especially) in the radically different future that we are creating with advanced science and technology.

## Cosmism versus Religion

I have a lot of issues with the institution of "religion," but I have to give it one thing: unlike academic philosophy, it excels at providing people with practical guidance on how to approach their lives and themselves.

But none of the religions around today are going to be of much use as advanced technology unfolds. Heaven above and hell below are going to seem increasingly irrelevant as uploading, human-level AI, brain-computer interfaces and molecular assemblers unfold.

Cosmism is the practical philosophy I try to use to approach my own life and self — and intend to use to face the very different situations that I may confront in the future — and my point in writing this Manifesto is to share this practical philosophy with others, in a simple and explicit way.

Cosmism may seem an eccentric bundle of ideas right now — but the relevance of the Cosmist perspective will become evident to more and more individuals as the next years, decades and centuries unfold.

## Some Cosmist Principles

If my take on Cosmism could be fully summarized in a list of bullet points, I wouldn't write a whole manifesto about it — I'd just write a few bullet points.

But, even so, it seems worthwhile to start with a few bullet points, just to whet your appetite for the more thorough and useful exposition to come.

Some of these bullets are rather abstract and initially may come across fairly opaque. That is a risk of compressing things into bullet-point form. Read the full text of the Manifesto, think on it a bit, and hopefully you will see that all these ideas have simple, practical, everyday meanings.

The ten basic "Prisco Principles" I listed above are almost obvious to anyone of the "right" cast of mind. The principles I will list below are meatier, and not everyone who considers themselves a Cosmist will accept all of them! Maybe nobody except "early 21st century Ben Goertzel" will ever accept all of them!

There is no litmus test for Cosmism. These are no more and no less than some principles that are interesting and important to me, and seem close to the heart of Cosmism.

And so, without further ado, some Cosmist principles:

1) **Panpsychism**: There is a meaningful sense in which everything that exists has a form of "awareness" — or at least "proto-awareness", as some would have it. In Peirce's terms, "Matter is mind hide-bound with habit."

2) **The Universal Mind**: There is quite likely some meaningful sense in which the "universe as a whole" (an unclear concept!) has a form of awareness, though we humans likely cannot appreciate the nature of this awareness very thoroughly, any more than a bacterium can fully appreciate the nature of human awareness even as it resides in the human body

3) **Patternism**: One often-useful way to model the universe is as a collection of patterns, wherein each entity that exists is recognized by some agent as a pattern in some other entity (or set of entities)

4) **Polyphonic reality**: The notion of an "objective reality" is sometimes useful, but very often a more useful model of the universe is as a collection of

overlapping, interpenetrating and intercreating subjective realities

5) **Tendency to Take Habits**: The universe appears to possess the property that, when patterns exist, they tend to continue ... much more than would be expected in a hypothetical random universe

6) **Compassion** is a critical principle of the universe, and is fundamentally an aspect of the Tendency to Take Habits (it's the spread of love and feeling from one mind to the next). Caring for other sentient beings (and if panpsychism is accepted, everything has a little bit of sentience!) is a critical aspect of evolving to the next levels beyond current human awareness and reality

7) Feeling and displaying compassion is important to the inner health and balance of a mind, as well as to the health and balance of the portion of the universe that mind is embedded in

8) **Causation is not a fundamental aspect** of the universe, but rather a tool used by minds to model portions of the universe

9) **Deliberative, reflective consciousness** is the specific form of "universal awareness" that arises in certain complex systems capable of advanced cognition

10) Goals are generally best understood, not as things that systems "have", but as tools for modeling what systems do. So, what goals a mind explicitly adopts is one question, but what goals the person is actually implicitly pursuing is often a more interesting question.

11) **"Free will" is not "free"** in the sense that people often consider it to be, yet there is a meaningful sense of agency and "natural autonomy" attached to entities in the universe, going beyond scientific distinctions of randomness versus determinism

12) **Science is a powerful but limited tool:** it is based on finite sets of finite-precision observations, and hence cannot be expected to explain the whole universe, at least not with out the help of auxiliary non-scientific assumptions.

13) **Mathematics is a powerful but limited tool**: it helps explicate your assumptions but doesn't tell you what these assumptions should be

14) **Language is a powerful but limited tool**: by its nature, consisting of finite combinations of tokens drawn from a finite alphabet, it may not be powerful enough to convey everything that exists in the mind of the communicator

15) **The human "self" is a cognitive construct** lacking the sort of fundamental reality that it habitually ascribes to itself

16) Society and culture provide us with most of what makes up our selves and our knowledge and our creativity — but they also constrain us, often forcing a stultifying conformity. Ongoingly struggling with this dialectic is a critical aspect of the modern variant of the "self" construct.

17) **There is no ideal human society** given the constraints and habits of human brains. But as technology develops further, along with it will come the means to avoid many of the "discontents" that have arisen with civilization

18) Humans are more generally intelligent and more diversely and richly experience-capable than the animals from which they evolved; but it seems likely that **we will create other sorts of minds whose intelligence and experience goes vastly beyond ours**

19) It seems likely that any real-world general intelligence is going to have some form of emotions. But human emotions are particularly primitive and difficult to control, compared to the emotions that future minds are likely to have.

**Gaining greater control over emotions** is an important step in moving toward transhuman stages of evolution.

20) **It is not necessary to abandon family, sex, money, work, raspberry-flavored dark chocolate** and all the other rewarding aspects of human life in order to move effectively toward transhumanity. However, it is desirable to engage in these things reflectively, carefully making a conscious as well as unconscious balance between one's need to be human and one's need to transcend humanity

21) Various tools like **meditation and psychedelic drugs** may be helpful in transcending habitual thought patterns, bringing novel insights, and palliating problems connected with the limitations of constructs like self, will and reflective awareness. But they do not fully liberate the human mind from the restrictions imposed by human brain architecture. Future technologies may have the power to do so.

22) Whether the "laws" and nature of the universe can ever be comprehensively understood is unknown. But it seems wildly improbably that we humans are now anywhere remotely near a complete understanding

23) Whether or not transhuman minds now exist in the universe, or have ever existed in the universe in the past, current evidence suggests it will be possible to create them — in effect to *build "gods"*

24) As well as building gods, it may be possible to *become* "gods." But this raises deep questions regarding how much, or how fast, a human mind can evolve without losing its fundamental sense of humanity or its individual identity

25) As we set about transforming ourselves and our world using advanced technology, many basic values are worth keeping in mind. Three of the more critical ones are **Joy, Growth and Choice** ... interpreted not only as personal goals, but also as goals for other sentient beings and for the Cosmos.

26) When confronted with difficult situations in which the right path is unclear, a powerful approach is to **obsolete the dilemma**: use a change in technology or perspective to redefine the reality within which the dilemma exists. This may lead to new and different dilemmas, which is a natural aspect of the universe's growth process.

27) Battles with the "enemies" of Cosmism are probably not the best path to achieve Cosmist goals. The universe is richly interconnected and

"Us versus Them" is often more realistically considered as "We versus Us." Struggles, including violent ones, are part of the natural order and can't be avoided entirely ... but there are often other ways, sometimes less obvious to the human mind; and part of the Cosmist quest is to find mutually beneficial ways of moving forward.

## Pattern Space

Preliminaries done, we now begin our journey into the particulars of Cosmism.

I will begin at a very abstract level — giving Cosmist answers to some of the good old Big Questions about life, the universe and everything. But if such philosophical explorations are not to your taste, don't worry — this is a short work, so in just a few pages, we'll be on to sex, drugs, uploading, superhuman AI, the future of society and a bunch of other juicy specifics. This Manifesto actually started out dealing only with those more specific topics, but as I got into the writing, I found that in order to discuss them in a coherent and unified way, it was necessary to give some deeper conceptual foundations first.

Let's get started with some small issues like

*Where do we live? What are we made of?*

*Is there some sort of underlying, universal Cosmos in which both our minds and our bodies, and the various things we see and experience, reside?*

*Is the physical world we move around in "real" in any absolute sense? Or is it "just" some sort of simulation — could we be living in a video game written by an alien computer programmer? (perhaps an amateurish, botched job produced by a young programming student?)*

Of course questions like these are useful more for stimulating thought than for attracting definitive and final answers. These are issues minds will likely be exploring for as long as minds exist.

One interesting way of addressing the Big Questions is to begin from the perspective that: we live in a world of patterns.

There are disorganized, teeming stimuli/forms/feelings of various sorts ... and then there are patterns organizing these entities.

A pattern, most generically, is something that brings simplicity to a complex collection of entities ... it's a "representation as something simpler"

The physical world presents itself to the mind, in large part, as a collection of patterns.

The mind presents itself to itself, in large part, as a collection of patterns.

One often-useful way to model the world we live and exist in is as a pattern space.

Of course, there is nothing "objective" about a pattern though ... whether a given entity is a pattern in some set of other entities is really a matter that's up to the perceiving mind.

In other words: a pattern is a representation as something simpler ... but who or what judges the simplicity?

The Cosmos is a space of interrelating entities, and individual minds are distinguished in large part by which entities they perceive as simpler than others, which largely determines which entities it sees as being patterns in which others.

This may seem a very abstract perspective, but it actually sheds a lot of light on various situations we encounter in our practical lives.

Approach the various things you encounter and experience, in the world and in yourself, as if they're patterns of organization — because that is largely what they are.

Many of the difficulties we have in life arise from unconsciously assuming that various things we encounter have some fictitious sort of absolute reality, going beyond the reality of pattern space.

"Everything is pattern" is not a panacea for cosmic understanding — I will introduce some further dimensions a little later — but it's a powerful perspective.

# The Tendency to Take Habits

David Hume was the first Western philosopher to point out really clearly what is now called the "problem of induction" — i.e. how is it we can assume that, just because the sun has risen every morning for the last thousand (or million) days, it will do so again tomorrow?

We can say it will continue because we have an intuition that patterns tend to continue — but then how do we know that patterns tend to continue? We know because in the past we have observed that patterns tend to continue. But then how do we know that this past observation, which is itself a sort of pattern, will tend to continue?

Charles Peirce used the term "the tendency to take habits" to refer to the assumption that, in the universe we live in, patterns *do* tend to continue.

An implication is that our ability to predict the future is predicated on our ability to identify patterns.

And, digging a little deeper — remember that pattern, in the view I've suggested above, is defined in terms of an *assumed simplicity measure.* This suggests that: For a mind to operate effectively, it needs to assume a simplicity measure for which, in its experience, the tendency to take habits holds true ...

We have the largely happy circumstance of living in a world where there are simplifications apparent from considering the past and the future together — where we can observe coherent patterns unfolding through time. We take this for granted but it's an important observation: without that, experience would be a lot of noise ... and intelligence and self would be impossible.

# First, Second, Third

Viewing the world as a system of patterns yields copious insights — but patterns aren't the whole story.

Patterns are relationships of a particular sort: a pattern is a relationship between one entity and a set of others, where the first is judged to represent and simplify the others.

The American philosopher Charles Peirce placed the "universe as a web of patternment relationships" perspective in a broader context by introducing the basic philosophical categories called First, Second and Third.

- **First** = pure, unprocessed Being

- **Second** = reaction ... the raw feeling of one thing having impact on another

- **Third** = relationship (the raw material for pattern: patternment is a particular, critical kind of relationship)

One can also push further than Peirce did, following other thinkers like Jung and Buckminster Fuller, and posit categories like

- **Fourth** = synergy ... networks of relationships spawning new relationships

In this perspective the view of Cosmos as pattern-space is the perspective of Thirdness.

# Isn't This Just a Bunch of Abstract Nonsense?

This sort of abstract categorization of Cosmos doesn't do much in itself ... but it provides a general perspective that can be useful for addressing concrete issues.

We will use this categorial perspective to approach the topics of awareness and consciousness ... which are critical to various issues that will confront us as technology develops, such as immortality, AI and uploading.

Philosophy pursued in the absence of practical issues tends to become verbal or intellectual gamesmanship.

Practical issues pursued in the absence of appropriate philosophy tend to get tangled in confusion — which can be fine; but given the sensitivity of the point in human history we're approaching, serious attempts at confusion-minimization seem indicated!

# Awareness is Everywhere

The idea that some form of "awareness" is everywhere, pervading everything, was considered obvious by the preponderance of pre-civilized cultures, and is now considered obvious by most practitioners of Buddhist meditation and many other wisdom traditions.

However, modern Western culture has led to a world-view in which most of the universe is viewed as somehow "dead" while only certain particular systems are viewed as having "awareness."

This new view of awareness has led to all manner of conceptual problems which philosophers enjoy debating. But, panpsychism — the old view that acknowledges awareness everywhere — remains the only view of awareness that is not plagued by complex contradictions ... as well as being an obvious truth to intelligences in appropriate states of consciousness.

Clearly, there are differences between the manifestations of awareness in a rock, a human, a society, an ecosystem and a universe — and these differences are worth attention and study. But we mustn't lose track of the universality and commonality of awareness.

## Awareness as First

In Peircean terms, "raw awareness" is First. Saying that everything is aware is saying that everything can be viewed from the perspective of First.

When we view things from the perspective of Thirdness, relationship, then the difference between humans and rocks seems dramatic and significant. We humans host far more complex pattern-networks than rocks.

Yet from the First perspective, we're all just sparks of raw awareness — people, rocks, equations, aliens from the 9th dimension... whatever.

## Universal Mind

If awareness is immanent in everything, is it not immanent in the Cosmos as a whole, as well?

Have we then reconstructed "God" within Cosmism?

This is a trick question, of course. "God" means many things to many people.

Some might say that I have found a strange path to "God," by way of Cosmism. Others would disagree, considering these philosophical thoughts unrelated to God or religious truth.

One thing is clear: The nature of the awareness of the Cosmos remains largely unknown to us measly humans. Just as, in many religions, the nature of the "mind of God" is considered beyond human ken.

We may feel we have a sense of it, in certain states of awareness — but then how can we truly know if we are connecting with the awareness of the whole, or just the awareness of a region of the Cosmos much larger and more powerful than ourselves, yet miniscule compared with the whole thing?

Fortunately we don't need to answer this question. It is enough to expand our boundaries, to connect with mind that goes beyond us and to become more than we could previously understand or imagine.

We may approach the awareness of the Cosmos incrementally through our ongoing growth, whether or not we ever get there (or whether "getting there" means anything).

## Patterns All the Way Down!

You've probably heard the story...

*The Eastern guru, holy lice in his beard and all, affirms that the earth is supported on the back of a tiger.*

*When asked what supports the tiger, he says it stands upon an elephant.*

*When asked what supports the elephant he says it is a giant turtle.*

*When asked, finally, what supports the giant turtle, he is briefly taken aback ... but quickly replies "Ah, after that it's turtles all the way down."*

(According my son Zebulon, after a few billion turtles, one comes to an enchilada, and after that it's enchiladas all the way down — but that's the same basic idea.)

The Peircean view of the universe as Third is much like this amusing parable.

In the realm of Third, it's patterns all the way down!

There are patterns ... but these patterns must be patterns of arrangement of *something,* of some "substrate."

But of what does this substrate consist? This substrate only enters into the realm of Third insofar as it presents itself as a set of observed patterns.

But, patterns in what?

These must be patterns in some substrate.

Or is it patterns all the way down?!

In the realm of science, the recursion ends with observations that are commonly accepted within some community. Some Master Dataset is accepted by a community as constituting valid observations, and then patterns are recognized in this dataset.

But the recognition of these commonly accepted observations as patterns in individual sense-data — this is where the "patterns all the way down" bottomless recursion is pushed, in the case of science.

You never reach a solid reality whose existence is just known ... not in the realm of Third. Because the only thing that is known is relationship, and a relationship must be a relationship among some entities. But if these entities are part of the realm of Third, then ... etc.

First and Second, in a sense, might seem to bottom out the recursion — but they don't really, because they are a different order of being.

The nature of life is that we just keep discovering relations among relations among relations ... and relations beneath relations beneath relations ... and never get to a fundamental reality underneath. We may get to things that *seem* fundamental — but how can we ever really know?


This fundamental bottomlessness of the realm of pattern and relationship underlies a lot of other issues that will occur a little later in these pages, including the possibility that our reality is in some sense a "simulation", and the notion of the universe as a multiverse or multi-multi-...-multiverse.

Turtles on turtles on ... turtles on turtles; patterns in patterns ... in patterns in patterns.

Mathematically one can only model this kind of recursion using obscure constructs like hypersets.

But experientially they are not hard to appreciate, if one avoids being caught up in "naive realist" perspectives that hold there is some absolute reality bottoming out the hierarchy/heterarchy of relationships; or "naïve nihilist" positions that maintain nothing has any meaning.

# Is Our World a "Simulation"?

Now let us shift, step by step, from general Cosmist philosophy to Cosmist views of coming technologies....

Anyone exposed to modern science fiction movies or novels must have asked themselves: Could we be living in a "simulation" world, like the world in the film The Matrix?

*Of course we could be.*

But one thing to notice about the situation in that movie is: the only reason it made any difference to the characters that their world was a simulation, was that there was *a way out* ... into some other reality beyond the simulation, in which the simulation could be viewed *as a simulation*.

In fact, the concept of "simulation" is too limiting.... A "simulation" is a copy of something else, and that's not really what we're worried about when we talk about our world being a simulation. What is really at issue is whether our universe is a *pliable, manipulable, adjustable external system* **from someone else's perspective**. Whether its *apparently fundamental* properties could be changed by some entity who lives outside it. In the rest of this chapter I will use "simulation" to mean "pliable, manipulable, adjustable external system for someone."

To say our world is a "simulation" in the commonly-used sense is, in essence, to say: there is another perspective from which the patterns that we observe as invariably true (that constitute our "objective world") are in fact pliable and manipulable, able to be changed around in various ways. And this must not be a purely theoretical perspective; it must be possible for some intelligent being to make use of this manipulability by changing around the nature of our world, causing it to become a different sort of world.

In principle, there is no way for us to know whether our world is a simulation or not. And, estimating the odds is mainly an exercise in futility, given our current level of knowledge.

But, it is not unreasonable to expect that becoming more intelligent and/or exploring more and more of our universe, could enable us to understand this possibility better — and maybe, to perform some Matrix-like "level-jump" into some perspective from which our world is a manipulable, pliable simulation.

But what are the odds that the perspective from which our world is a simulation is very similar to our own world?

This is also hard to answer. Maybe the perspective or "universe" from which our world is a simulation is wholly different from our own perspectives as residents of *this* universe. We might not even recognize it as a "world" at all, if we were to encounter it.

Just as the realm of Third is patterns all the way down, the universe may well be simulations all the way down ... but this is only interesting if there is some practical way to access the simulations/realities at levels below ours in the posited hierarchy....

# The Pattern of the Individual Intelligent Mind

If one views the world as patterns among patterns among patterns ... then each of us, as individual minds, must be viewed as a pattern as well!

What kinds of patterns are we?

There are many ways to answer this question; here I'll give one answer that has proved useful to me in my science and engineering work designing AI systems and analyzing human mind/brains — and that seems to also tie in with what various wisdom traditions have said about the individual mind.

## What is an Individual Mind?

An individual "mind", from the view of Third, can be thought of as the set of patterns associated with some intelligent system.

And what is an intelligent system?

An intelligent system can be thought of as: A set of patterns that is capable working together as a "system," toward the purpose of achieving complex goals.

The broader the collection of goals the system can deal with, the more general its intelligence.

## Subjectivity of Intelligence

"Achievement of complex goals" might seem a profoundly limited conceptualization of intelligence, given the limited role that explicit goal-achievement plays in real intelligences.

But if one thinks in terms of *implicit goals* — the goals that a system looks to be working towards, based on what it is actually doing, regardless of how it conceptualizes itself — the perspective starts to seem more broadly applicable.

One arrives at the notion of an intelligent system as *one that can sensibly viewed or modeled as seeking to achieve complex (implicit) goals in complex environments, using limited resources.*

This emphasizes that intelligence is in the eye of the beholder — because it takes some beholder to assess what are a system's implicit goals.

What looks intelligent to A may not look intelligent to B, depending on what implicit goals A and B respectively are able to recognize.

This characterization of intelligence highlights the vast variety of intelligent systems that are possible — which derives from the vast variety of goals that may be pursued by different systems, in a vast variety of possible environments.

## Generality of Intelligence

We humans have a certain degree of general intelligence — but we are not *wholly* general minds. We are a wild mixture of general and specialized capability.

Our brain has limited capacity, so there are many things our brains — in their current forms, or anything similar — can never understand or do.

It seems unlikely that any *absolutely* general intelligence can ever be created using a finite amount of (computational or energetic) resources. Any finite system is going to have some biases to its intelligence — some goals and environments it does better on.

Much of human intelligence may be understood as adaptation to the specific bodies, goals and environments in which our minds evolved to operate — though as we advance culturally, psychologically and technologically we are progressively generalizing our intelligence.

But our brain also has the capability to expand itself by augmenting its "hardware" infrastructure — which means that, transhumanistically speaking, it's not so limited after all.

Given that we have the capability to flexibly self-modify, there are no clear limits to what we may become. Limits may be discovered as we progress. And, even if there are no limits to what we can become — there may well be limits on how generally intelligent we may be come and still be considered *human*.

## No Limits

Our individual minds may appear strictly limited — if we view them from certain limited perspectives (the "consensus" perspective of modern Western society, for example.)

But such a perspective is itself a construct of the individual mind: it is something the individual mind learns and builds for itself, just as the individual mind builds its understanding of the "external world."

The wisest perspective is one in which individual mind and external reality create each other. This is the only view that reconciles the inner experience of existing, with the apparent presence of entities like rocks and snowstorms that are difficult to morph via mere power of thought.

As John Lilly eloquently put it:

*"In the province of the mind,*

*in the inside reality,*

*what one believes to be true,*

*either is true or becomes true*

*within certain limits.*

*These limits are to be discovered*

*experimentally and experientially.*

*When so determined these limits are found to be*

*further beliefs to be transcended."*

## Mind, Body and World

The specific patterns that we are, as *human* minds, are intimately bound up with our human bodies, and with our ways of remembering and interacting with the world these bodies live in.

## We are Embodied Minds

As minds that are associated with particular physical systems, we are closely tied to the sensors and actuators of these systems.

Jack Kerouac described himself as "just another soul trapped in a body" — and I've often felt that way — but it's not the whole story.

We think with our hearts, lungs, digestive systems and genitals and so forth — not just with our brains. If you took a human brain and connected it to a different sort of body — or left it to cognize in a void with no body — it would fairly quickly self-organize into something radically different that would only marginally qualify as "human" or as "the same mind" that came before.

# Varieties of Memory and Cognition

We are minds devoted to sensing and acting in environments. In order to do this we remember the environments we've experienced. And importantly, we remember them in multiple ways:

- **sensorially,** including multi-sensory integration
- **episodically**: remembering the essence of experiences, even if we forget the sensory details
- **declaratively**: abstracting general facts, beliefs and ideas from masses of (largely forgotten) experiences
- **procedurally**: remembering how to do things, even if we don't exactly remember the why of all the steps we take
- **intentionally**: we remember how we broke our goals into subgoals in various situations
- **attentionally**: we remember what sorts of things merited our attention

Each of these kinds of memory constitutes a different kind of pattern, and is associated with different kinds of dynamics for pattern recognition, formation, and combination.

For example, declarative memory naturally ties in with reasoning

- procedural memory naturally ties in with what the psychologists and engineers call "reinforcement learning" — learning via getting reward and punishment signals, and automatically adjusting one's behavior accordingly
- sensory memory (especially visual memory) naturally ties in with hierarchical structures for pattern recognition.

The human brain contains particular intelligent pattern manipulation dynamics corresponding to each memory type — and AI systems may contain different dynamics serving similar purposes, with different strengths and/or weaknesses.

How much of this sort of humanlike brain architecture is specific to humanlike minds, and how much is characteristic of minds-in-general, is something we are still discovering.

Other sorts of minds, like those of cetaceans or distributed Internet intelligences, will likely still have memory and processing functions corresponding to the categories mentioned above — but will likely carry out each of these functions very differently!

# The Phenomenal Self

What is this thing called "self"? — this inner image of "Ben Goertzel" that I carry around with me (that, in a sense, constitutes "me"), that I use to guide my actions and inferences and structure my memories?

It is nothing more or less than *a habitual pattern of organization in the collection of patterns that is my mind* ...

... which is correlated with certain *habitual patterns of organization in the collection of patterns that is the mind of the portion of the physical and social environment I habitually associate with.*

My "self" keeps telling itself that it is the mind associated with my body ... and in trying to make this story true, it usually succeeds to some degree of approximation (though rarely as high a degree as it thinks it does!) ... but ultimately it is *not* the mind associated with my body, it is just a portion of that mind which has some overall similarities to the whole.

Thomas Metzinger, in his wonderful "neurophilosophy" book *Being No One*, uses the intriguing term "phenomenal self" ...

Seeing the self as the self-constructed dynamical phenomenon it is, is one of the main insights that commonly results from meditation practice or psychedelic drug use.

The attachment of primal awareness to self is part of what characterizes our deliberative, reflective consciousness.

Self wishes and acts to preserve itself — this is part of its nature ... and is also part of the intense aversion many humans feel toward death, and the intense drive some humans feel for immortality.

If the *whole mind* wants to be immortal, it will be *partly* satisfied by spawning children, writing books, and so forth — things that extend the patterns constituting it further through time. (Woody Allen's quip "I don't want to be immortal through my works — I want to be immortal through not dying" notwithstanding!)

If the *self* wants to be immortal, it doesn't really care much about offspring or literary works — it just wants to keep churning along as a self-creating, self-persisting dynamical subsystem of the mind.

It is unclear the extent to which transhuman minds will have "selves" in the sense that we humans do. Part of "human selfness" seems to be an absurd overestimation, on the part of the self, of the degree to which the self approximates the whole mind. If this overestimation were eliminated, it's not clear how much of "human selfness" would be left. Some of us will likely find out ... ( — although, the issue of whether it will be "us" that finds out, or some descendant of us, is part of the question at hand!)

## The Extended Self

Each self-model — each phenomenal self — has its own holism and integrity, yet also stands in containment and overlapping relations with a host of other selves.

An individual person has a localized self — and also an extended self, which includes various aspects of the "inanimate" world they interact with, and also aspects of various people they interact with.

If you put me on a deserted island for 10 years, a lot of my self would disappear — not just because it needs stimulation from others, but because my extended self actually resides collectively in my brain and body, in the brains and bodies of others, and in the environments I and these others habitually inhabit.

This is/was obvious to many other cultures, and feels slightly odd to some of us now only because we live in an unprecedentedly individualistic culture.

## Both Objective and Subjective

So which is it – are we

- **minds generated by physical bodies,** living in a real physical world

or

- **minds that generate "body" and "physical world"** as part of our thought-activity, and part of our coupling with other minds in the overall creative mind-field

??

Trick question, of course: It's not either/or!

Both perspectives are sensible and important, and Cosmism embraces both.

Mind is part of world; world is part of mind. Subjectively, each of us in-some-sense "builds" the world ourselves, from our sense-perceptions — but yet, isn't it amazing how this world we "build" takes on so many properties that we didn't explicitly put in? Clearly, although there's a sense in which the world is something my mind builds up from its perceptions, there is also something going on in the world beyond my individual self and what it could possibly make. It feels more accurate to say that there is some pattern or possibility field "out there", and my mind's activity uses its sense-perceptions as a seed out of which the world crystallizes, drawing in material from this field as it goes.

Objectively, on the other hand, each of our unique experience-streams seems to emerge from dynamics in particular hunks of physical tissue. Tweak the brain a bit, with a scalpel or a screwdriver or merely a little red pill, and the mind changes radically.

There's no contradiction between these two views. We can use them together.

Mathematically, one can model this sort of "circular creation" using structures called hypersets. But one doesn't need fancy math to understand what's going on — one simply needs to look openly at reality and experience, and not try to impose any particular perspective as primary.

# Causality (A Convenient Construct)

We humans like to think in terms of causality ... but causality seems not to be an intrinsic aspect of the universe.

Rather, causality is something we impose on the universe so as to model it for various practical purposes. We do this both consciously and unconsciously.

## Causality is Not Scientific

No currently accepted scientific theory makes use of the notion of causality. Scientists may interpret some math equations involved in a scientific theory to denote causality — but unlike, say, "force" or "attraction", causality is not really part of the formal language of modern science.

Roughly, causality consists of "predictive implication, plus assumption of a causal mechanism."

Predictive implications are part of science: science can tell us "If X happens, then expect Y to happen with a certain probability." But science cannot tell us whether X is the "cause" of Y, versus them both habitually being part of some overall coordinated process.

## Causality and Will

Our psychological use of causality is closely related to the feeling we have of "free will." Understanding causality as a construct leads quickly to understanding "free will" as a construct. The two constructs reinforce and help define each other.

On a psychological level, "X causes Y" often means something like "If I imagined myself in the position of X, then I could choose to have Y happen or not to have Y happen." So our intuition for causation often depends on our intuition for will.

On the other hand, the feeling of willing X to happen, is tied in with the feeling that there is some mental action (the "willing") which *causes* X to happen.

Will and causation are part of the same psychological complex. Which is a productive and helpful complex in many cases — but is also founded on a generally unjustified assumption of the "willing or causing system" *being cut off from the universe.*

## Our Minds are Enmeshed in the Cosmos

Cosmism accepts that individual minds are embedded in the Cosmos, enmeshed in complex systems of influence they cannot fully understand. This means a mind cannot really tell if a given event is causal or not, even if that event occurs within that mind.

So assumptions of causation or willing may be useful tools in some context for some purposes — but should be understood as pragmatic assumptions rather than objective, factual observations.

# Natural Autonomy:  Beyond the Illusion of Will

Nietzsche said that free will is like an army commander who takes responsibility, after the fact, for the actions of his troops.

Modern brain science has proved him remarkably on-target: Gazzaniga's split-brain experiments, Benjamin Libet's work on the neuroscience of " spontaneous" activity and a lot of other data shows that when we feel like we're making a free spontaneous decision, very often there's an unconscious brain process that has already "made the decision" beforehand.

## So Are We All Just Automata?

So what does this mean? That we're all just automata, deterministically doing what the physics of our brains tells us, while deluding ourselves it's the result of some kind of mystical spontaneous conscious willing?

Not exactly.

Science's capability to model the universe is wonderful yet limited. Contemporary science's models of the universe in terms of deterministic and stochastic systems are not the universe itself, they're just the best models we have right now. (And these days they're not even a complete, consistent model of our current set of observations, since general relativity and quantum theory aren't unified!)

The evidence clearly shows that, when we feel our "willed decisions" are *distinct, separate and detached* from our unconscious dynamics, we're often at variance with neurophysiological reality.

But this does not imply that we're deterministic automata....

It does imply that we're more enmeshed in the universe than we generally realize — specifically: that our deliberative, reflective consciousness is more enmeshed with our unconscious dynamics than we generally realize.

## Intentionality Beyond the Illusion of Will

Might there be some meaningful sense of **intentional action** that doesn't equate with naive "free will"?

Yes, certainly.

But this meaningful sense of "intentional action" must encompass the enmeshed, complexly nonlinearly coupled nature of the mind and world.

I.e., it's not intentional action on the part of the deliberative, reflective consciousness as a detached system.

It's "intentional" action on the part of the Cosmos, or a large hunk thereof, manifested in a way that focuses on one mind's deliberative, reflective consciousness (perhaps among other focii).

The "intentionality" involved then boils down to particular kinds of patternment in a sequence of actions.

For example — among other aspects — "choice-like" action-sequences tend to involve reductions of uncertainty — reductions of "entropy" one might say ... collapses of wide ranges of options into narrower ranges.

When our deliberatively, reflectively conscious components play a focal role in an appropriately-patterned entropy-reducing dynamic in our local hunk of the Cosmos, we feel like we're enacting "free will."

## Natural Autonomy

Henrik Walter, in his book *The Neurophilosophy of Free Will*, develops some related ideas in a wonderfully clear way.

He decomposes the intuitive notion of free will into three aspects:

1.  Freedom: being able to do otherwise
2.  Intelligibility: being able to understand the reasons for one's actions
3.  Agency: being the originator of one's actions

He argues, as many others have done, that there is no way to salvage the three of these in their obvious forms, that is consistent with known physics and neuroscience. And he then argues for a notion of "natural autonomy," which replaces the first and third of these aspects with weaker things, but has the advantage of being compatible with known science.

He argues that *"we possess natural autonomy when*

1.  *under very similar circumstances we could also do other than what we do (because of the chaotic nature of the brain)*
2.  *this choice is understandable (intelligible — it is determined by past events, by immediate adaptation processes in the brain, and partially by our linguistically formed environment)*

3.  *it is authentic (when through reflection loops with emotional adjustments we can identify with that action)"*

The way I think about this is that, in natural autonomy as opposed to free will,

- Being able to do otherwise is replaced with: being able to do otherwise in *very similar* circumstances
- Agency is replaced with: emotionally identifying one's phenomenal self as closely dynamically coupled with the action

Another way to phrase this is: if an action is something that

- depends sensitively on our internals, in the sense that slight variations in the environment or our internals could cause us to do something significantly different
- we can at least roughly model and comprehend in a rational way, as a dynamical unfolding from precursors and environment into action was closely coupled with our holistic structure and dynamics, as modeled by our phenomenal self

then there is a sense in which "we own the action." And this sense of "ownership of an action" or "natural autonomy" is compatible with both classical and quantum physics, and with the known facts of neurobiology.

Perhaps "owning an action" can take the place of "willing an action" in the internal folk psychology of people who are not comfortable with the degree to which the classical notion of free will is illusory.

Another twist that Walter doesn't emphasize is that even actions which we do own, often

- depend with some statistical predictability upon our internals, in the sense that agents with very similar internals and environments to us, have a distinct but not necessarily overwhelming probabilistic bias to take similar actions to us

This is important for reasoning rationally about our own past and future actions — it means we can predict ourselves statistically even though we are naturally autonomous agents who own our own actions.

Free will is often closely tied with morality, and natural autonomy retains this. People who don't "take responsibility for their actions" in essence aren't accepting a close dynamical coupling between their phenomenal self and their actions. They aren't *owning their actions*, in the sense of natural autonomy — they are modeling themselves as *not* being naturally autonomous systems, but rather as systems whose actions are relatively uncoupled with their phenomenal self, and perhaps coupled with other external forces instead.

None of this is terribly shocking or revolutionary-sounding — but I think it's important nonetheless. What's important is that there are rational, sensible ways of thinking about ourselves and our decisions that don't require the illusion of free will, and also don't necessarily make us feel like meaningless, choiceless deterministic or stochastic automata.

## Shaping and Flowing

We humans evolved to be as smart as we are, not just because of our brains, but also because of our hands.

We like to make stuff with our hands. We like to pick up sticks and bash things down. We like to build things out of sticks and blocks.

We like causing and building.

All this is very well — but it provides dramatic habituation to our conceptual vocabulary.

We would do well to think a little less in terms of causing and building, and a little more in terms of shaping and flowing.

Instead of causing and willing things, we should more often think of ourselves as flowing along with broader processes with which we are correlated.

Instead of building things, we should more often think of ourselves as shaping and influencing ongoing processes.

This is not to say that shaping and flowing are invariably better concepts than causing and building. Just that we habitually overemphasize the latter and underemphasize the former. Other kinds of minds might have different biases.

These particular biases of ours go along with our general bias — which is to some extent a human bias, and to some extent a modern-Western-culture bias — to overemphasize our degree of separation from the Cosmos with which we're enmeshed.

# The Theater of Reflective, Deliberative Consciousness

In Orwell's Animal Farm, the ruling pigs famously change their slogan "All animals are equal" to "All animals are equal, but some animals are more equal than others."

Panpsychism accounts for the human experience of consciousness in a similar way: "All entities are aware, but some are more aware than others."

Or, just as much to the point: some are *differently* aware than others.

Every entity in the universe — every pattern — has some awareness, but each pattern manifests its awareness differently depending on its nature.

Our reflective, deliberative "theater of consciousness" is the way that primal awareness manifests itself in *one part of our mind/brain*.

As Bernard Baars has articulated nicely in his cognitive science work, this theater of consciousness integrates all the kinds of memory and processing that our minds do — it's the "place" where "it all comes together." (I surround "place" with quotes because in the case of the human brain it's not a physical location — it's an emergent dynamical pattern involving multiple regions, and different ones in different cases.)

According to panpsychism, the "unconscious" parts of your mind/brain are in fact "conscious" in their own ways — but their own less-intense consciousness is only loosely coupled with that of your theater of reflective, deliberative consciousness.

Various practices such as meditation or psychedelic drug use may increase this coupling, so that the reflective, deliberative consciousness can become more closely coupled with the consciousness of the other parts of the mind/brain that normally appear to it as "unconscious."

None of this however should be taken to deny the specialness of the theater of reflective, deliberative consciousness. It's a wonderful phenomenon — it's definitively, gloriously different than what takes place in rocks, atoms, molecules, clouds or even lizards. Puzzling out its structure and dynamics is an important task on which cognitive neuroscience is gradually making headway.

But, what makes this aspect of our minds special is *not* that it's the unique receptacle or source of awareness (it isn't ... nothing is).

## Consciousness and Separation

Part of what characterizes the theater of reflective, deliberative consciousness is the *special effort* it makes to decouple itself from the unconscious. To an extent, it cuts itself off from perceiving the awareness of the other parts of the mind/brain, so it can carry out processing using processes that ignore these other parts.

The reflective/deliberative consciousness wants to gather some information from the unconscious, and then process it in an isolated way, because that way it can carry out special processes that wouldn't work otherwise.

Reflective/deliberative consciousness works in part by making near-exhaustive intercombinations of the small number of things in its focus at any given time. It couldn't do this if it opened up its scope too much, due to the limited amount of resources at its disposal.

So we have a very important theme here: limitation of resources is causing a system (the reflective/deliberative consciousness) to increase its degree of separateness, so as to enable it to achieve some goals better within the resources at its disposal. But these goals themselves have to do with persisting separateness (in this case the separateness of the organism associated with the mind containing the reflective/deliberative consciousness). Separateness spawns more separateness.

Separateness often makes things more interesting ... and often also less joyful ... a general theme to which I will return later.

# Joy and Woe

Pleasure and pain as Firsts are, like all Firsts, raw and unanalyzable.

They simply are what they are.

Saying they are something else, is a matter of drawing relationships and patterns, and thus moves one into the realm of Third.

In the midst of a moment of pleasure or pain, analyses are irrelevant. The experience is what it is.

But from the point of view of planning our lives, relationship is important: we want to understand what is likely to bring joy or pain to ourselves or others, or to the world as a whole. And we want to understand what pleasure and pain are, in a relational sense. What kinds of organization-patterns are they?

I will use the terms **joy** and **woe** to refer to the complex mind-networks that arise corresponding to the raw feelings of pleasure and pain. Joy and woe are much more complex than pleasure and pain; among other factors, joy may involve some pain mixed with the pleasure, and woe may involve some pleasure mixed with the pain.

In spite of their subtleties, however, these are simpler emotion-patterns than many other ones. They are not specific to humans or any kind of organism — they are very generic patterns, infrastructural to the Cosmos.

## Joy as Increasing Unity

Paulhan, a psychologist writing 100 years ago, had the very interesting insight that "happiness is the feeling of increasing order."

Joy is unity. Joy is togetherness. Joy is the gaps getting filled, so that there's no more emptiness craving to be sated, but everything is newly filled-up and satisfied.

Joy is the feeling of increasing unity.

Of course there is much more to the human experience of joy than this — all human emotions are complex, multifaceted beasts. But this is part of it, and an important part: it's the pattern-space dynamic at the heart of joy.

Minds contain various expectations: meaning, they contain internal representations of patterns they would like to see emerge from their experience, and they seek out experiences that will cause these patterns to emerge. When the seeking ends and the pattern emerges, there is a feeling of unity: the mind and the experience are bound together. Patternment has increased.

A body feels joy when patternment that is central to its function and integrity increases.

A self feels joy when patternment that is central to its function and integrity increases.

A mind feels joy when patternment that is central to its function and integrity increases.

## What About Drugs?

What about the pleasure that comes from smoking crack, for example? Is this true joy?

It is, in fact, a brief rush of incredible unity. Everything flows together into one long joyous instant of raucous excitement, or boundless oceanic bliss, etc.

Unfortunately it's short-lived — and it's achieved by shutting down many important parts of the mind, so that the others can exist together in a unity of excitation.

During a crack high, parts of the mind are experiencing great joy, and other parts are effectively put to sleep.

Other drugs like psychedelics provide a more profound and holistic experience — but still there are almost always peculiarities, frustrating mixtures of insight and exaltation with self-delusion and confusion. The drug experience at its worst is dangerous and destructive, and its best is wonderful and enlightening yet still frustratingly flawed.

This is different from some kinds of natural joy, in which all the parts of the mind and body are bound together in one joyful experience.

As an example, those who achieve nirvana-type mind-states via psychedelic drug use may have an equally profound experience to those who achieve it via meditation or other non-drug means. But on average, enlightenments achieved via drugs are harder to sustain and harder to integrate with ongoing life, even though they  may be easier to come by in the first place.

## Joy and Growth

The role of **increase** in joy is worth reflecting on. We habituate quickly to new pleasures. Most of the time, lottery winners are ecstatic for a while, but in the long run are no happier than others.

Stasis is not the path to joy. This simple fact becomes important when considering the various pathways open to humans as technology advances. There is more potential joy in developmental trajectories that lead to a continuing onset of new unities, than trajectories founded on "more of the same."

## Woe and Pain

What about joy and pleasure's antonyms, woe and pain?

Pain, Paulhan notes, is associated with the feeling of decreasing order: disharmony, disunity, the dissolution of patternment.

Of course, there is much more than this to the human experience of pain, and the associated mind-complexes of woe — but, this is pain's pattern-space core.

From the perspective of a whole body, mind or self, joy and woe are not necessarily opposites — because these are all complex systems, and increasing disunity in one part can be coupled with increasing unity in another. And it is hard to say that raw pain and pleasure are opposites either, because *opposition* itself is a Third, out of place in the realm of pure experienced Firsts.

# Complexity of Human Joy and Woe

Joy and woe are simpler than other emotions, at core — but they also grow, within the human mind, into complex, coordinated systems.

There are neurological disorders which cause people to experience pain without the painful aspect. That is: they realize they are having a sensation identifiable as "pain," but there is no emotional component — the pain doesn't seem to hurt!

What this indicates is that the sensation of the body being damaged (or the mind being damaged, in the case of psychic pain) is not quite the same thing as the negative emotional experience attached to this sensation! Ordinary human pain is the result of a complex coordination between sensation and emotion (and often cognition), but in these brain damaged individuals, the coordination is broken.

And something similar can happen with joy. Certain kinds of depression involve *being happy but not enjoying it*. The positive evaluation is there, but not the emotional component that's normally attached to it.

When the abstract structure of joy (increasing unity) or woe (decreasing unity) is allied with the emotional centers, then we have the typical, glorious human experience of joy or woe.

## Emotion

Emotions play an extremely important role in human mental life – and perhaps because of this, we are sometimes fascinated by the idea of intelligence without emotion. Like Mr. Spock in the original Star Trek, or Mr. Data in the newer Star Trek — an alien and robot respectively, who are more intelligent than most humans yet lack the feeling-driven component of human behavior (Spock however is much more appealing and moving than Data, to most of us, precisely because he does have emotions, they're just generally repressed and not a controlling factor in his mind or life.)

But are emotionless intelligences really possible? To what extent is human emotion a consequence of our particular evolutionary heritage, and to what extent is it an aspect of Mind In General?

Clearly, much of human emotional life is distinctly *human* in nature, and not portable to systems without humanlike bodies. Furthermore, many problems in human psychology and society are caused by emotions run amok in various ways –  so in respects it might seem desirable to create emotion-free AIs one day.

But there are limits to the extent to which this will be possible. Emotions represent a critical part of mental process, and human emotions are merely one particular manifestation of a more general phenomenon –  which must be manifested in *some* way in any mind.

The basic phenomenon of emotion is something that will exist in any mind that models itself as having some form of "free will", and may be conceptualized as

*An emotion is a mental state that does not arise through a feeling of "will" or "autonomy", and is often accompanied by physiological changes*

Human emotions are elaborations of this general "emotion" phenomenon in a peculiarly human way.

There are a few universal emotions –  including joy and woe –  which any intelligent system with finite computational resources is bound to experience, to an extent. And then there are many species-specific emotions, which in the case of humans include rage, fear and lust and other related feelings.

Controlling our emotions is, by definition, never going to be fully possible. However one can certainly adopt a mental dynamic in which emotions are not the controlling factor — in which spontaneous emotional arisings are incorporated in a high-level ratiocination process, and thus don't affect action on their own. In other words: Spock is possible; Data probably not. Even a highly rational digital intelligence would probably be more Spock-like than Data-like, even if its flavors of emotion were less humanlike than Spock's.

## Compassion

We tend think about compassion on the level of individual selves and minds: Bob feels compassionate toward Jim because Jim lost his wife, or his wallet, etc. Bob sympathizes with Jim because he can internally, to a certain extent, "feel what Jim feels."

But it's often more useful to think of compassion on the level of *patterns*.

The pattern of "losing one's wife" exists in both Bob and Jim. Its instance in Bob and its instance in Jim have an intrinsic commonality — and when these two instances of the same pattern come to interact with each other, a certain amount of joy ensues ... a certain amount of increasing unity.

Compassion is about minds adopting dynamics that allow their internal emotional patterns to unify with other "external" patterns.

It is about individual minds not standing in the way of pattern-dynamics that seek unity and joy.

The tricky thing here is that individual minds want to retain their individuality and integrity — and if the patterns they contain grow too much unity with "outside" patterns, this isolated individuality may be threatened.

The dangers of too much compassion are well portrayed by Dostoevsky in *The Idiot,* via the tale of the protagonist Prince Myshkin — who goes nuts because of feeling too much compassion for various individuals with contradictory desires, needs, ideas and goals.

There seem to be limits to the amount of compassion that a mind can possess and still retain its individuality and integrity. However, it seems that (unlike Myshkin) mighty few humans are pushing up against these limits in their actual lives!

And of course, transhuman minds will likely be capable of greater compassion than human minds. If they have more robust methods of maintaining their own integrity, then they will be able to give their cognitive and emotional patterns more freedom in growing unity with external patterns.

## Should Compassion Be Maximized?

Should compassion be maximized? This is a subtle issue.

From the point of view of the individual, maximization of compassion would lead to the dissolution of the individual.

From the point of view of the Cosmos, maximization of compassion would cause a **huge burst of joy**, as all the patterns inside various minds gained cross-mind unity.

But would the joy last? Joy is about *increase* of patternment. An interesting question about this hypothetical scenario of maximal compassion is: After every mind wholly opened up to every other mind and experienced this huge burst of compassion, would there still be a situation where new patterns and new unities would get created?

Perhaps some level of noncompassionateness — some level of separation and disunity — is needed in order to create a situation where new patterns can grow, so that the "unity gain" innate to joy can occur?

# The Balancing Act of Human Compassion

We should be compassionate. We should open ourselves up to the world.

We should do this as much as we can without losing the internal unities that allow our minds to operate, to generate new patterns and new unities.

And we should seek to expand and strengthen our selves so as to enable ever-greater compassion.

Our selves and our theaters of reflective, deliberative consciousness are frustrating and even self-deluding in some regards — but they are part of our mind architecture, they are part of what makes us *us*. At this stage in our development, they are what let us grow and generate new patterns. We can't get rid of them thoroughly without giving up our humanity, without sacrificing ourselves in a sudden and traumatic way.

Perhaps, as transhumanist technology advances, many of us will choose to give up our humanity, via various routes. Perhaps in doing so we will achieve greater levels of compassion and joy than any human can. But until that time, we have to play the dialectical game of allowing ourselves as much joy and compassion as we can, while keeping our selves and our internal conscious theaters intact enough to allow us to function in our human domain.

This may sound like a frustrating conclusion, but the fact is that nearly no one pushes this limit. Quite surely, outside of fiction I've met very few individuals who experience so much compassion it impairs their ability to function!

## Postscript by Samantha Atkins

Upon reading an earlier version of the above in the preliminary online version of this book, Samantha Atkins commented as follows, reiterating the ideas based on her own experiences, and integrating several other themes touched on elsewhere in this Manifesto:

*Everyone that has meditated or done certain drugs or just followed certain paths of reflection has dissolved the self in "Self", lost self in transpersonal pattern. It is only scary when you are paranoid that "you" will cease or not arise again. Everyone coming out of a psychedelic experience has watched the "self" reintegrate out of the seeming cosmic "not-self" — a place where "self" doesn't seem relevant or even believable. You can even give a tweak here and there to "self" as it reconstructs. So what are we? Good question.*

*...*

*I think the dance from self to Self and back again, concurrently at different points of the cycle, is a very much more realistic consciousness. It is neither easy or hard to achieve. It is hard to maintain and function adequately in all settings where only self is expected. Big compassion changes you fast. Even opening to just compassion/oneness/equal importance of a small group of people changes you a lot. There is a reason people that do that much go off to special places; and if they want to do it all the time they tend to stay there.*

*...*

*One thing I hope and suspect is that we learn that the limit on our own wealth/happiness/wellbeing is the asymptote of the maximal actualization of the highest potentials of all others. After all, our selves all dance with, enliven, enrich, add value to our shared space and one another. Thus the maximization of all those others is the maximization of ourselves. This is hard to see within scarcity based thinking. But I think it is essential to see to ever really experience abundance, no matter how much we have.*

...

*I understand how to go to that level of compassion, all connection that impairs function. I have touched it, dipped into it, been attracted, been repelled, found it hard to keep an even keel in the everyday world. Mostly I was not willing to let go to the changes I felt happening and required to live there. Someday I may decide differently.*

## Joy, Growth and Choice

What general values can we identify as important, beyond culture-specific or species-specific or otherwise context-specific moral codes or ethical values?

To put the question another way ... an earlier version of this Manifesto began with the definition

*Cosmism: a practical philosophy centered on the effort to live one's life in a positive way, based on ongoingly, actively increasing one's understanding of the universe in its multiple aspects*

Later this got modified into

*Cosmism: a practical philosophy focused on enthusiastically and thoroughly exploring, understanding and enjoying the Cosmos, in its inner, outer and social aspects*

*Cosmism advocates*

- *pursuing joy, growth and freedom for oneself and all beings*
- *ongoingly, actively seeking to better understand the universe in its multiple aspects, from a variety of perspectives*
- *taking nothing as axiomatic and accepting all ideas, beliefs and habits as open to revision based on thought, dialogue and experience*

One difference is that when I wrote the latter I decided to specify what I mean by "a positive way." I.e., I decided to get a little more concrete about the critical question of: what are the important values?

There are many Cosmist values, and it would be folly to attempt a definitive enumeration.

However, as reflected in the above proclamation, three values seem particularly essential to me: Joy, Growth, and Choice.

I've discussed these above, but will now revisit them from a "value-system" perspective.

**Growth** is perhaps the simplest: the creation of new patterns, out of old ones.

I don't want stasis. Nor degeneration. Some old patterns may need to cede to the new, but overall there should be an ongoing flowering of more and more new patterns.

Note that growth is not just the constant appearance of new patterns — it implies some **continuity**, in which old patterns are expanded and improved, yielding new ones that go beyond them.

**Joy** I have analyzed as the feeling of increasing unity, togetherness, order.

We want more and more new patterns to be created and we want them to get bound together into unities and wholes, to have the joy of coming together.

We want there to be minds that can experience this joy — the joy of coming together with their environments and each other.

**Choice** is the most complicated of the three values I've identified — but it's also the only one that implies the existence of integral, individual minds in anything like the sense that humans have them.

You could have growth and joy in a Cosmos without individuals — but choice requires individual minds. Valuing choice means valuing individuals that decide. These individuals don't need to have illusions of all-powerful, unpredictable free will — they may well be more realistic about understanding their choices as being associations between their internal dynamics and broader entropy-reducing dynamics in their region of the Cosmos. But still they have their own intentionality.

Compassion, in this view, comes down to valuing joy, growth and choice in a way that goes beyond the boundaries of one's individual mind, body or self.

A Cosmos of individuals, choosing their actions and experiencing joy, growing in a joyful growing Cosmos — this is close to being the crux of what Cosmism values. *Living life in a positive way: living life in a way that promotes and embodies universal joy, growth and choice.*

## Love

There may be no more confusingly, gloriously polysemous English word than "love."

Love of your country, love of your family, love of chocolate, of a favorite movie ... love of life ... passionate romantic sexual love ... love of God; pure love with no object ...

What does it mean?

Love has to do with empathy, but it's not the same — you can empathize a lot with those you only love a little; and sometimes people display shockingly little empathy for those they love deeply...

According to the Beatles, "all you need is love" — but according to Charles Bukowski, "love is a dog from hell."

What most of the varieties of "love" have in common is: Love is an emotion a mind has toward someone or something else, that is associated with experiences going *beyond the self.*

And if one wants to get metaphysical, then there's "universal love" — the love of the Cosmos for itself! Whenever one part of the Cosmos is separated from another, there's a chance for those two parts to overcome their separateness via love.

Some of the particular varieties of love we humans experience may be inapplicable to transhuman minds — and these love-specifics may change a great deal even for humans, as we modify and advance ourselves.

Minds without sexual reproduction and death wouldn't have the same kind of family love as we do — let alone the complex beautiful mess of romance....

Don't necessarily expect Valentine's Day to continue past the Singularity!

But love itself — emotion toward an Other, which brings us beyond our Self — which extends our Self to encompass that Other — love will almost surely survive, in any future that embraces joy, growth and choice. And not "just" universal love — love between individual minds will almost surely exist, as long as individual minds do. Because minds do seek joy, growth and choice — and love is an extraordinarily powerful way for a mind to grow beyond its self, experiencing a joy greater than it could have on its own....

## Obsolete the Dilemma!

It's all very well to enunciate lovely-sounding values like Joy, Growth and Choice ... but in real life we're all faced with difficult decisions. We're faced with choosing one being's joy over another's, or choosing joy versus growth in a given situation, and so forth.

There's no perfect, one-size-fits-all solution to such dilemmas.

But Cosmism does provide one valuable principle, that is very frequently appropriate for beings in the phase of evolution that humans currently occupy.

This is the principle of *obsoleting the dilemma.*

Rather than trying to resolve the dilemma, use a change in technology or perspective to redefine the reality within which the dilemma exists.

This may of course lead to new and different dilemmas — which is a natural aspect of the universe's growth process.

This approach has tremendous power and we'll revisit it frequently in the following pages.

To make the idea clear, first of all I'll explore it in the context of a couple simple, everyday issues that — in the human world right now — seem to have a tremendous power to divide thoughtful, compassionate people.

Cosmism doesn't solve these issues — but it does advocate a systematic route to resolving them ... not by solving them but rather by *obsoleting* them.

## The Dilemma of Abortion

Abortion is one of the most obvious cases of a divisive ethical dilemma, in modern society.

Even among individuals who reject traditional religious notions of the human soul and the special sacredness of human life, it poses a huge ethical challenge.

On the one hand, compassion dictates that killing babies is wrong ... and the fact is that we don't really know when a fetus develops enough "reflective awareness" that killing it becomes more like killing a person than like killing a sheep, or more like killing a sheep than like killing a fish, etc.

On the other hand, forcing an adult human woman to create an infant when they don't wish to, is a clearly uncompassionate violation of that woman's personal choice and happiness, of her ability to grow in the directions she chooses.

So, there is a balance to be struck, and different caring, thoughtful people want to strike it in different ways.

## The Dilemma of Vegetarianism

Vegetarianism presents a similar dilemma to abortion, though one that the mainstream of modern society seems less concerned about (due to our habitual species-centrism).

Clearly, killing animals to eat them is uncompassionate and, in itself, "wrong" according to principles of joy, growth and choice. Cows, pigs and chickens may not be as smart as we humans are — but they have their own experiences, emotions and wills, and we're pretty damn nasty to abort these so we can have a tastier dinner.

Yes, nature is bloody and violent ... animals kill each other ... but, compared to nonhuman animals, we have the capability to much more fully understand what we're doing and to make a more considered choice....

Yet there are plenty of borderline cases. It's not clear, for instance, whether fish experience pain in the same way that birds and mammals do. It's not clear in what sense a fish has a theater of reflective consciousness. Personally, I don't feel confident that killing a fish is cutting off a stream of tremendous joy and experience, any more so than cutting down a tree or picking a carrot out of the ground.

One argument against vegetarianism is that we're evolved to be omnivores and some level of meat consumption is necessary for us to feel "natural." I personally do feel that way: in the past when I've eaten a vegetarian diet for a while, I have felt a certain disturbing lack of energy and aggressive initiative. Eating fish cures that for me just as well as eating other meat ... but without *some* fish or other meat intake, I really don't feel like "myself." So the question becomes: how much cruelty to fish would I incur to gain a certain amount of personal energy?

After all, we're also evolved to kill each other when we get mad — but we make a point of suppressing this evolutionary urge in the interest of our mutual growth, joy and choice, etc.

## Obsoleting the Dilemmas

What does Cosmism have to contribute to these familiar dilemmas?

It doesn't provide any trick for drawing the line between right and wrong in these tricky situations. Joy, growth and choice and other Cosmist principles are all surely relevant; but these dilemmas are cases where two or more different options exist, each bringing joy/growth/choice and other goodies to some minds at the cost of others ... and so there's a difficult judgment to be made.

What Cosmism suggests is an alternate path: ***obsolete the dilemma.***

This is already happening, to some extent. We should try to make it happen far more as the future unfolds.

Birth control largely obsoletes the dilemma of abortion, though it doesn't quite work well enough yet. The ability to remove an embryo from the mother without pain or danger, and incubate it in a lab from a very early stage, would obsolete the abortion dilemma in a different way.

The capability to grow cloned steaks, fish cutlets and chicken breasts and such in the lab would obsolete the dilemma of vegetarianism ... as would advances in synthetic food technology; or pharmacology that conferred the recreational, physiological and neurochemical benefits of various forms of food without requiring actual food ingestion.

There is a powerful general principle here. Ethical dilemmas are never going to be completely avoidable, but the advance of technology can blunt them pretty thoroughly, if it's done with a specific eye toward obsoleting the dilemmas.

## The Dilemma of Poverty and Charity

The issue of poverty and charity can be perceived in much the same way as abortion and vegetarianism.. The ethical dilemma of whether to send 80% of my income to help starving children in Africa (I never do so, but feel some guilt over this), will be neatly obsoleted by advanced technology that eliminates material scarcity.

Why don't I send 80% of my income to help starving children in Africa (nor even 10%, for that matter)? Due to the usual mixed motivations. Part of it is surely plain old selfishness; I don't claim to be a wholly altruistic individual. And part of it is a sense that the world as a whole would *not* be better off if those of us fortunate enough to live in wealthy nations (or the upper classes of poorer nations) were to revert to the economic mean.

If I sent most of my income to help starving children I would be less happy on a day by day basis — but also, I would also be in much less of a mental and practical position to create new technologies, Cosmist Manifestos, and so forth ... things that I value a great deal. My own children would be in a much worse position to create such things as well, if I were to deprive them of books, computers and education so as to feed the starving kids in Africa. And yet, I'm never quite sure it's right to value these creations over peoples' lives.

What I'm quite sure of, is that it's right to obsolete the dilemma.

## Dialectics Redux?

If you've had some contact with Marxist or Hegelian philosophy you may find something familiar in the "obsolete the dilemma" notion.

Hegel, as a key point of his philosophy, described how thesis and antithesis lead to synthesis via the "dialectical" process. The synthesis obsoletes the dilemma between the thesis and antithesis. The dilemma between Being and Nothingness is resolved to yield Becoming. The dilemma between lords and vassals is resolved to yield new social classes emerging due to the advance of industrial technology. Marx saw the advance of society as a result of a series of dialectical dilemma-resolutions.

In Hegelian dialectics, dilemmas are obsoleted by redefining realities so that previously oppositional realities become unified in a new set of structures and dynamics. This is indeed closely related to the Cosmist notion of obsoleting the dilemma (and arguing the precise relationship would be an onerous task of technical philosophy that I won't undertake here!).

However, one big difference between the Hegelian/Marxist perspective and the Cosmist perspective is the amount of determinism that the former perceives to exist in the world. The notion of a precise, orderly series of dilemmas, getting obsoleted and then leading to new dilemmas in a predictable fashion — this is anathema to the Cosmist perspective, which is all about embracing the unknown and growing oneself so as to understand and become new things that would have been wholly incomprehensible to one's prior self. Often, once a previous dilemma has become obsoleted, the world looks like a totally new place ... and the path forward is one that you never could have imagined to exist before.

Society has not evolved according to anything like the particular path that Marx and Hegel foresaw. It *has* evolved according to a "quasi-dialectical" process of iterative dilemma-obsoletion, though ... and will continue to do so ... so those guys did get *some things* fundamentally right.

Figuring out how to obsolete the dilemmas facing us is an ongoing intellectual, emotional, social and spiritual challenge — not at all a matter of following some predetermined and inevitable path.

## Wisdom

If intelligence is about achieving complex goals — what is wisdom?

Wisdom has many aspects....

One of them is effectively leveraging experience (episodic knowledge) toward intelligence.

Another is breadth — generality of intelligence, rather than narrow-focus on particular sorts of complex goals.

And another has to do with what sorts of goals to focus on. Wisdom has to do with focusing on greater, broader goals — broader in extent (beyond the individual self, encompassing other individual minds and even the Cosmos), and broader in temporal extent (long-term goals rather than short-term goals).

So one view of wisdom is:

*Being able to use experience, along with other mental aspects, to understand and achieve a broad variety of things, with a focus on broader and longer-*

*term things rather than self-focused or immediate things...*

One interesting consequence of this view of wisdom is: it makes clear that seeking to go beyond the human condition as presently conceived, may be the best way to achieve greater wisdom....

So if one says that a strategy like "obsoleting the dilemma" is wise, this means that it is useful for pursuing wisdom — for understanding and achieving a broad variety of things, with a focus on broader and longer term things rather than self-focused or immediate things....

And more broadly: The basic goals of Cosmism can be seen as a natural route to the achievement of increased wisdom....

## Respecting Ecosystems

One of the major dilemmas that needs to be obsoleted as we move forward is that between technology and environmental preservation.

Ecosystems are beautiful, complex things — in fact, they are highly intelligent minds, though apparently lacking theaters of reflective, deliberative consciousness.

Further, ecosystem-minds blend richly into human minds ... we have evolved to include animals, plants and systems thereof in our extended minds. Many of us find we think and feel much more clearly when in the forest or other natural surroundings — and there's no strange mystery here; it's because our extended minds encompass aspects of nature.

The advent of industrial and advanced technology is causing great damage to the ecosystem of Earth, leading to conflicts between some of those who deeply love the advance of technology and some of those who deeply love the Earth.

However, I think this is a temporary, albeit tragic, phenomenon. As technology advances further it will incur less and less environmental destruction, not more and more.

In fact, we can already see this happening.

Technologically advanced nations are not the worst environmental offenders — those are developing nations, who use cruder technologies more egregiously inflicting of environmental damage. Wealthier, more advanced nations value their clean air and water and parks to a greater extent.

The dilemma of "technology and environment" will be obsoleted as technology develops further, and becomes more flexible and efficient in its use of resources.

# The Sociocultural Mind

One of the charming peculiarities of modern Western culture — and especially American culture, in which I've lived imost of my life, and which has played a pivotal role in the development of humanity's advanced technologies — is its emphasis on the individual rather than the social group.

And yet, if you took a human infant and raised them isolated from culture, what would they be? A primitive being, not so different from an ape. As has been repeatedly demonstrated, without appropriate social interaction in early childhood, humans never develop full linguistic and cognitive abilitites. And in most cases, people cut off from social interaction for more than a few years wind up going mad.

The truth is that human society and culture are a vast meta-mind with greater computational power, insight and complexity than any individual human mind.

So far as we can tell, the emergent mind of human society currently lacks the level of global coherence that an individual human mind has. There's no theater of reflective social consciousness. But even if so, human society contains a huge number of complex high-level self-organizing patterns, that course through all us individuals, and guide us in multiple ways, many of which our selves and deliberative consciousnesses are barely or un aware.

## The Inescapability of Culture

Technological and cultural progress have largely been driven by individuals who deviate from social norms — who push beyond their society and culture, often suffering greatly for this.

But yet, the directions that they're pushing in, the ideas and movements they're pushing forward, are invariably defined in terms of human culture and society. We, as humans, simply don't know anything else.

Chinese, aboriginal and American culture (to pluck three random examples) may seem wildly different to us — but of course, in the bigger space of all possible modes of sociocultural interaction, they're really mightily mutually similar.

It's possible that some radically different culture and set of mind-states could be implemented in a set of interacting human brains and bodies ... but we have no way of discovering that, because a culture is not something one person can build on their own, nor something that a group of people can simply get together and spontaneously create. It's far bigger and more complex and subtle than that.

## Rebellion as Conformity

In the end the existence of rebellious individuals pushing beyond the norms is "just" a mechanism that modern society and culture uses to grow, extend and expand itself. Those of us who act as rebels, are following the deep self-organizing patterns of the sociocultural order just as thoroughly as those who act as conformists. This wouldn't have been quite so true for some wild-eyed rebel emerging in a steady-state culture like aboriginal culture ... but it's very true now, given that modern culture relies on ongoing rebellion to generate the constant stream of scientific, cultural and aesthetic advance on which it's predicated.

Rational, orderly-looking growth curves like the ones Ray Kurzweil draws, projecting the advances of various technologies, already factor in the likelihood of a large number of individuals risking (and in many cases suffering) ostracism to spend their lives promoting ideas that their peers consider moronic, maniacal and/or misguided.

# Will Posthumans Have Use for Friendship?

One fascinating question regarding the future of society is whether the traditional phenomenon of "friends and family" will persist among posthumans.

Various complexities may arise such as the emergence of mindplexes (minds with theaters of reflective consciousness at the social scale as well as the individual scale) ... but even so, it seems plausible that each individual mind will choose to couple itself especially closely with a small number of other individual minds, for extensive information-sharing.

One reason for this might be that, even with "quasi-telepathic" knowledge sharing between minds, it may take some effort to really get to understand another mind well ... so that once one has undertaken that effort, it makes sense to continue the coupling with that mind, so as to reap the rewards of the understanding one has gained via sharing that mind's detailed thoughts and experiences.

On the other hand, maybe the ability to share thoughts more directly than is possible among humans will obsolete the need for close ongoing relationships with particular individuals — perhaps this new ability will enable close rapport to be established between any two reasonably compatible minds within a brief period.

It all depends on how well transhuman forms of thought-sharing really work, given the innate incommensurability of different minds' views of the world.

## Sociocultural Self-transcendence

The modern sociocultural mind has a pattern of violating and transcending itself.

That is: it creates dilemmas, then advances in such a way as to obsolete them ... and in the process it creates new dilemmas, etc.

While Hegel and Marx's sociology made a lot of errors, it did get this general pattern of "dialectical advance" about right. They saw how society advances to obsolete dilemmas — e.g. the dilemma of feudal lords versus vassals was obsoleted via the industrial revolution. Which in turn led to new dilemmas to be obsoleted.

Cosmism itself exemplifies this self-violating/self-transcending aspect of the modern sociocultural mind: it both violates and emerges from the dominating patterns in modern society and culture.

And it seems likely that this overall pattern is going to lead to the ultimate and thorough transcendence of the modern sociocultural mind — as it pushes us humans (who compose its "cells") to fundamentally modify or replace our physical substrates ... a move that will dramatically transform the nature of the sociocultural mind we compose.

The individual mind and the socio-mind will transcend together.

The dilemmas of "individual versus society" and "mind versus body", which lie at the root of many of our contemporary social and psychological problems, are likely to soon get obsoleted — and replaced with other, more advanced dilemmas! As a part of the natural growth of the thought process of the social mind with which we all, as individuals, interpenetrate.

# How Should Society Be Structured?

Democracy ... capitalism ... communism ... socialism ... anarchism ... the list goes on and on ...

Is there any really good way to structure a human society?

If not, then what's the best of the bad lot?

Or, to put it another way: Is there some way that a sociocultural mind can be fashioned out of human "cells", that leads to rampant joy, growth and choice ... and with less of the opposites of these values than we now see in the human world around us?

Certainly, Cosmist values seem to argue in favor of societies allowing their members a fair bit of personal freedom, and encouraging progress rather than a steady state ... and treating each other with compassion rather than cruelty or repression.

But there are difficult issues when one digs into the details, and broad Cosmist philosophy doesn't solve them.

Balancing compassion versus choice in government is difficult — taxing people to pay to feed and educate poor children is an imposition on freedom; yet leaving the poor children starving and ignorant is uncompassionate. One would like society to self-organize in such a way that such dichotomies don't exist — and perhaps this is happening, coupled with the advance of technology; but it's happening frustratingly slowly for those now in disadvantaged positions.

It seems clear that, even without further advanced technologies, we could do considerably better than current social orders as solving these difficult problems.

But ultimately, it seems there are limits to how well a society of humans can be structured , given our intrinsic cognitive limitations. And there are even stricter limits to how well a society of humans can be structured *under conditions of scarce resources*.

I.e.: any sociocultural mind, composed at base of "legacy humans" like we are today, is bound to be at least *a bit* psychologically screwed-up ... and even more so if we're involved in struggling over scraps of matter shaped into particular configurations.

The best course, it seems, is to obsolete the dilemmas of society by redefining the human mind and abolishing material scarcity.

# Improving Democracy?

Churchill said "Democracy is the worst form of government, except for all the others."

Clearly there's something to this, which is why the democracy meme is spreading around the world, and is correlated with other indices of human well-being. Although, there is some evidence that representative democracy works significantly better in societies that have surpassed a certain minimal level of education and wealth.

But even if the value of democracy is accepted: what kind of democracy, then?

Should the Net be used to allow direct democracy? — where people vote on a variety of specific issues rather than delegating so much to their representatives (who so often seem to confuse and corrupt matters, sometimes doing so even in spite of good intentions)? But who has time to study the details of complex legislative issues so as to vote on them intelligently?

Presumably new mixtures of direct and representative democracy will emerge as advanced technology more thoroughly permeates our culture. To an extent — though an incomplete extent, giving the limitations of human mind

— Internet technology can likely obsolete some of the dilemmas of democracy.

# Human Life After Scarcity?

One thing that could improve the human political situation a lot more than tweaking the details of democracy, would be the drastic reduction of material scarcity.

The main reason human life is less brutal now, by and large, than in the past, is because advances in technology have reduced gradually scarcity. We have developed various useful social institutions that embody a great deal of calculation and wisdom ... but arguably, the main reason these institutions are proving workable and stable, is the concurrent technological advance, which has come hand-in-hand with advances in education and wealth.

It seems quite possible that democracy is only stably achievable under conditions of relative material abundance. The ancient Greeks achieved democracy's practical preconditions via slavery. In the modern world we've achieved them via technology.

Most likely, resources will never be infinite — there will always be some contention for resources of some sort.

Most probably some form of scarcity will always exist. But relative to our innate human desires, scarcity could dwindle close to zero. Given nanotech molecular assemblers plus ultra-realistic virtual worlds, provided free of charge to all, there would be a lot less motivation for anyone to risk their comfort by fighting over resources.

In this sort of scenario, the task of managing a society, using ever more refined and liberating mechanisms, will become far easier.

All sorts of possibilities that would seem "utopic" from the present perspective may become possible. Many of the social dilemmas that now strike us as inevitable, may be drastically obsoleted.

# A Post-Scarcity Network of Enclaves?

For instance: Perhaps in a post-scarcity scenario, it would work for small enclaves of humans to form — communities centered around different belief systems and life-patterns.

Each enclave could basically have its own little world, and there wouldn't be practical competition between enclaves due to the minimization of scarcity.

In the modern situation, this sort of "enclave" based social arrangement would run into problems due to issues like pollution that span enclaves. But it seems feasible that advanced technology could resolve this ... "all" one needs are extremely inexpensive nonpolluting molecular assemblers, for example.

You might argue that this is a fanciful notion, because scarcity will never truly be eliminated. But, it may be possible to eliminate scarcity from the perspective of everyday human life. There's a limit to how much an individual human can consume, and still remain human.

What if scarcity were reduced sufficiently that no one wants for practical physical comforts ... and there is more than enough advanced media and entertainment and intellectual and artistic technology to go around: more than anyone could possibly use in thousands of years? In this sort of regime, it seems quite possible that the urge to invade other enclaves and take their resources would be extremely rare.

## Better Societies through Better Brains

More pessimistically, there is the possibility that we humans are collectively *so* perverse that we will still battle each other viciously even once scarcity is virtually abolished.

If so, then the only solution to making a happy society — to *really* obsoleting the social dilemmas — is to modify the human mind/brain ... solving the problems of society at the source.

There is no doubt that, if the human mind/brain were dramatically improved, avenues for deeper social interaction and cultural invention would open up, making modern societies seem more obsolete than primitive tribal life seems today.

I.e., in this scenario: The emergent sociocultural mind would become a far more growing, joyful, purposeful "individual".

## Practical Politics Today

The above thoughts on society may seem utopian, unrealistic, futuristic ... what about the fray of real-world politics, right now?

Cosmism dictates only a few broad principles in this regard ... and in practice, reconciling them may be difficult! For instance,

- Allow development of advanced technologies except in cases of extreme danger
- Extend compassionate help to all, except where the imposition on individuals is tremendous
- Don't build structures or dynamics in stone — leave each aspect of society free for adaptation and growth
- Make joy, growth and choice explicit goals of the social order

Broad principles like these manifest themselves in many ways in the particular situations we now confront — but that would lead us beyond the present text, to a different sort of manifesto!

# Sexuality & Beyond

One futurist friend of mine likes to tell people of his aspiration to somehow remove all sexuality from his brain.

Most people think he's nuts in this regard (including his wife, with whom he has an active and healthy sex life), but he's convinced that sexuality is one of the main factors slowing down our progress toward the Singularity, immortality, superhuman benevolent AIs and all the other good stuff.

One point he always makes in these conversations it that it's amazing, when you really think about it, how much of modern human society is structured around sexuality.

Marriage, kids, dating ... buying nice clothes and making oneself up to impress the opposite sex ... buying cars or houses or the latest cellphone to impress the opposite sex with one's success ... etc.

Sexuality pervades nearly everything in our lives, implicitly even when not explicitly.

As is now common knowledge, the power sex has over us is rooted in the power our DNA has over us. We are evolved to obsess over reproducing, over extending our

DNA to future generations. Even though most humans in First World countries now use birth control for nearly all their sexual encounters, and many humans choose not to reproduce at all, we are still strikingly controlled by the mind-patterns ensuing from our DNA's urge to persist itself.

But evolution has tangled sex up rather thoroughly with other aspects of our psyches. As Freud, Reich and others pointed out so thoroughly, human motivation is deeply tied with our inner sexual energy. Eunuchs seem to generally lack aggressive, enthusiastic motivation even for things outside the realm of sex. But when my anti-sex futurist friend speaks of blotting out sexuality from his mind, he doesn't want to blot out his passion and energy generically — he wants to focus it on things other than simulations or instantiations of the reproductive act.

And yet — for all the distraction that it provides — sex, at its best, is a profound "altered state of consciousness" experience, just as self-melting and reality-changing as meditation or psychedelics or any other extreme of human experience. It brings you outside of your ordinary self and state-of-consciousness, putting you in touch with wholly different ways of being, interacting, experiencing.

## Beyond Sex

A couple decades ago there was a color glossy periodical called "Future Sex Magazine." It didn't survive too long, partly because there wasn't that much to say on the topic. Teledildonics is amusing to read about once but in the absence of effective available technology the notion loses novelty fast ... and apparently there was a limited audience for monthly photos of hot models making it with robots and electronic AI-powered uber-vibrators.

What's more interesting is the prospect of new forms of experience, providing the same things sex provides — but also going *beyond sex* in significant, surprising ways.

As satisfying as sex and sexuality are for human bodies and minds, there seems little doubt that, eventually, transhuman minds will discover new forms of pleasure and fulfillment going far beyond what we now get from sexuality. Sexuality is amazingly wonderful but it's also a *mess* ... it wraps up confusion and pain with pleasure in complex ways, surely more than is necessary based on the innate interconnection of pain and pleasure that exists in any finite mind.

It may seem cold or eccentric to say so, but the fact is that orgasms and genitals and romantic relationships — as glorious as they are — are hideously badly designed. It's obvious that other entities better serving the same purposes (and more) could be developed...!

But what do we do now, while we're stuck in legacy human form? Should we embrace asceticism, so as to more effectively work toward transhumanity without distraction? Or should we make the most of the pleasures afforded by the human form, while they're the only ones we have ... so long as we can do so without killing our chances to find something even better later on?

To what extent is a healthy sex life needed to give the human mind/body an effective grounding for making the difficult judgments that will come in the next decades as technology develops?

There are no easy and immediate answers to these questions ... each of us has to make our own judgment.

All Cosmism does is urge us to make such judgments based on a rich understanding of the issues from multiple perspectives. And urge us to *obsolete the dilemma*, by fixing the underlying problem, which in this case is the rootedness of our experience in a body and brain that are very difficult for us to reflectively modify and manipulate according to our deliberative desires and conclusions.

# What Would Asexual Aliens Think of Human Sexuality?

One interesting exercise is to view sexuality as if one were an asexual alien viewing humans objectively and for the first time. What would such an alien think about the role of sexuality in human society and psychology?

Assuming the alien perceived the human race as being on the cusp of a technological Singularity or something similar, what would the alien think about the optimal human relationship with sexuality and other in-built obsessions of the human brain/mind?

## The Essence of Sex

Considering the potentially broad scope of alien sex brings us to a more fundamental question: What is the "abstract essence" of sexuality, apart from its particular manifestation in our human embodiments?

Is it just, say, *"the mutual exchange between two intelligences of positive emotion and pleasure-center stimulation, coupled with emotional open-ness?"*

That sort of abstract sex seems like an unabashedly wonderful thing ... but of course, there is so much (good and bad) about human sexuality that isn't implied by such an abstract essence.... For one thing such a notion of sex is uncoupled from reproduction. It is also not intrinsically tied to any notion of long-term commitment — although there may be an implicit relationship there, since emotion-sharing may work best between two minds that know each other well, in the way that only long-term togetherness can bring.

Probably there are forms of experience that embody the abstract essence of sexuality — and intensify it more than is possible within the constraints of humanity — and without the downsides of human sexuality (though perhaps with new downsides that we can't now imagine).

Bring 'em on!

# Goals and Meta-Goals

Ever have the experience that you seriously think you're trying to achieve one thing, but then in hindsight, years later, you look back and feel like your past self was actually trying to achieve something else entirely?

I.e., you weren't really after what you thought you were after?

I've had that experience myself on occasion — for example, during part of the time I was involved with my first startup company, Webmind Inc. I thought I was chasing two goals: 1) making a lot of money, and 2) getting advanced AI built. In hindsight though, often the actions I was taking weren't balancing these two goals very well. Much of the time I was straightforwardly pursuing AI research — and fooling myself that the things I was doing were really serving the money goal as well, to a greater extent than was really the case. Of course this was easy to do at that time, because it was the late-1990s dot-com boom, and no one really knew the secret to making money in that period anyway.

My *explicit goals* with that company were making money and getting advanced AI built, with roughly equal weighting. That's what I thought I was doing, at the time.

My *implicit* goals, the goals it looks in hindsight like my actions were pursuing, were weighted a bit differently: maybe 70% AI and 30% money. That's what I now think I was doing.

In regard to that company, at that point in time, I had a *poorly-aligned goal system*.

These concepts have a fundamental importance.

As an external observer, one can look at a system and identify the goals it seems to be pursuing: i.e., in mathematical terms, the functions it seems to be trying to maximize. These are the system's *implicit goals*.

A system can pursue implicit goals, in this sense, even if it lacks any concept of what a "goal" is.

Some systems, on the other hand, also have *explicit goals,* meaning that they model themselves as pursuing certain goals.

A *well-aligned goal system* is a set of **explicit goals** that fairly accurately reflect the **implicit goals** of the intelligent system containing the goal system.

Achieving a well-aligned goal system generally requires long practice and deep self-understanding.

Naturally, this is much easier if one's (explicit and implicit) goals are simple ones!

## Meta-Goals

*Well-alignedness* is an example of a *meta-goal*: a property of a goal system that is not tied to the specific content of the goals, but rather to the general nature of the goal system.

Another meta-goal is *consistency*.

Consider a goal system as a set of top-level goals, together with other subgoals that are derived from these. Then, a goal system is *fundamentally inconsistent* to the extent that achieving any one of the top-level goals, intrinsically decreases the level of achievement of any of the other top-level goals. Fundamental consistency is the opposite of this.

There is also a notion of *subgoal inconsistency*: the extent to which, on average, achieving the known subgoals derived from a particular top-level goal, decreases the level of achievement of the known subgoals derived from other top-level goals. For instance, it may be that the goal of having children is not fundamentally inconsistent with the goal of making scientific progress — but all the specific methods one knows for dealing with children are inconsistent with the goal of making scientific progress, so the two goals are subgoal-inconsistent even though not fundamentally inconsistent. Resolving this kind of situation can be done with effort, but the effort may be substantial.

Even if a system is fundamentally consistent, if it's not subgoal consistent, it may have a very hard time achieving its goals.

A related problem is *subgoal alienation* — sometimes a subgoal is derived as a way of achieving some other goal or subgoal, but then the former one persists even after the latter one is abandoned. Subgoal alienation leads to the accidental creation of new top-level goals, which leads to inconsistency and poor goal system alignment.

I've known a number of people who originally took high-paid, unrewarding jobs for the reason of saving money up to pursue some other goal. But after a while the other goal became less and less important to them, and the high-paying job became an end in itself. Sometimes this is subgoal alienation; sometimes it's a matter of a poorly-aligned goal system (maybe the high pay was really their main goal all along, and they were just fooling themselves about the other goal).

Sexuality is a massive case of subgoal alienation, on the cultural and biological level. It emerged as a subgoal of reproduction, but — especially since the advent of birth control — it has liberated itself from these roots and now serves as a top-level goal itself for most adult humans.

## How Goal-Oriented are Humans?

Humans, by nature, are not that thoroughly goal-oriented. We have certain in-built biological goals but we've done well at subverting these. Some of us adopt our own invented or culturally-acquired goals and put a lot of effort into pursuing them. But a lot of our behavior is just plain spontaneous and not terribly goal-oriented.

Of course pretty much any behavior could be modeled as goal-oriented; i.e. every series of actions can be viewed as an attempt to maximize some quantity. But the question is whether this is a simple and natural way to model the behavior in question — does it pass the Occam's Razor test?

# Will Advanced AIs Be Goal-Oriented?

One can imagine artificial minds with a vastly more thorough goal-orientation than humans, and a much greater attention to meta-goals. Most likely such minds will exist one day — alongside, potentially, artificial minds that are much *less* goal oriented, much more loosely organized, than humans.

It also seems likely that, even in a mind more thoroughly devoted to goal-achievement than the human mind, a certain percentage of resources will wind up being allocated to spontaneous activity that is not explicitly goal-oriented in any of its details ... because this spontaneous activity may generate creative ideas that will be helpful for achievement of various goals.

# Evolution of Top-Level Goals

I've spoken about about goal systems as possessing "top level goals," from which other goals are derived as subgoals.

Are these top-level goals then invariant as a mind evolves and learns?

This is not how humans work, and needs not be how AIs work.

Indeed, if one has a mind that gradually increments its processing power, one might expect that as it grows more intelligent it will discover new potential top-level goals that its earlier versions would not have been able to understand.

How can top-level goals get revised?

Not in a goal-oriented way, because if goal G is put in charge of revising goal G1, this really just means that goal G1 is not top-level — G is top-level.

Top-level goals can get revised via spontaneous, non-goal-oriented activity — and the occurrence of this phenomenon in intelligent systems seems to be a fundamental aspect of the growth of the Cosmos.

Goals grow, goals drift — and thus the universe evolves, via both pursuit and spontaneous development of goals.

# What (If Anything) Should My Goals Be?

Haha — don't expect Cosmism to tell you … at least not my flavor of Cosmism.

Nietzsche's version of the prophet Zarathustra said "*Go away from me and resist Zarathustra*!" — and I had a certain affinity for this perspective, long before I named my first son Zarathustra.

My favored flavor of Cosmism advocates the principles of growth, choice and joy as top-level goals.

But regarding how to interpret these — and how to break them down into subgoals — it advocates diversity, not conformity.

By working out the details on our own and with others — and hence making our own choices — we can find joy and promote growth for ourselves and the Cosmos.

## Work and Play

One of the many things I learned from reading Hannah Arendt is the value of carefully distinguishing *work* from *labor*.

*Labor*: the ongoing exertion of effort required to keep a living organism effectively functioning. Can bring great joy ... and pain ... is not intrinsically oriented toward growth. Will be increasingly obsoleted as technology advances.

*Work*: the creation of *works* ... the exertion of effort to make new things (which may be material, conceptual, social-relational, etc.) ... oriented toward growth as well as joy

The mixture of work with labor characterizes the modern era, but the correlation of the two will decrease as technology advances.

Once work is separated from labor, what remains? Essentially, work as art-work and social communion ... the creation of scientific, mathematical, visual, engineering, architectural works, not to put food on the table, but to gain social relationships and most of all *just for the feeling of doing it and the joy of getting it done*.

*Social action*: the creation of works whose impact lies in the social realm ...

Arendt's book *The Human Condition* gives a rather clear and erudite exposition of the above categories.

She also makes the bold assertion that *only through social action are people able to truly express freedom, and able to truly be human.*

If one interprets "being human" as "contributing substantially to the collective, emergent **mind of human society**" then she is correct.

# Is Labor Necessary?

Work and social action seem critical for advancing Cosmist values ... labor less and less so as technology advances

Whether the human body needs some sort of labor to be joyful is another question — but if it does, one may view this as a shortcoming of the human body-mind rather than as an indication of the fundamental cosmic importance of labor.

According to what we know of physics, some entity must "labor" in some sense in order for physical dynamics to happen ... for metabolism to occur, for structures to get fabricated, etc.

But technology has the capability to push more and more of the labor onto entities with less and less intense sentience, away from entities with rich theaters of deliberative awareness and high levels of intelligence.

As our minds, society and technology advance, work and social action should become increasingly disssociated with labor — among humans and other intensely, deliberatively conscious beings we may create or evolve into.

If this doesn't happen, it will probably mean we are handling our technological transcension in some profoundly wrong way.

## The Power ... and Limits ... of Play

And what of work's sometime antonym, play?

*Play*: not just spontaneous joy-inducing activity ... much of play involves the pursuit of goals that are analogous to, but easier to achieve than, goals an intelligent system finds important. Via pursuing these analogous goals, the organism may learn something about how to achieve the real goals of interest. The joy of play comes from the intrinsic activity, but also from the analogical connection to important goals...

If you're forever playing, and only playing, then play loses much of its meaning, which comes from its analogy to real-life goals.

Children can fully enjoy a life of pure play, because evolution has crafted their psychology to be "that of folks who will become adults." But it's famous how fast pure play grows boring for most early retirees.

For grown-ups, alternating play and work/social-action, with rich analogies growing and changing and binding the two realms of activity, is probably the most fulfilling way to live.

One could engineer a mind to enjoy an endless life-story of pure play without any need for work or social action. No doubt some human minds have self-organized into such a condition, already. But Cosmism views this as suboptimal: pure play will never lead to powerful growth. And without growth, ultimately, the scope of joy is limited — part of growth is the ability to experience more and more joy as one expands the scope of one's capabilities and experiences.

All work and no play surely does make Jack a dull boy — but (recalling the distinction we've drawn between work and labor) so does all play and no work.

## The End of Education

As civilization has advanced, education has become increasingly important — and increasingly pervasive.

This trend is going to continue, until "education" as a separate category dies, replaced *for those who choose to grow* by learning that thoroughly pervades life.

## Education at the Dawn of the Internet

Young (and not so young) people spend more and more of their lives in school ... and further, education is increasingly a regular and ongoing part of a person's career. The pace of technological and social change is such that it's rarer and rarer for a person to receive, in their youth, all the training they'll need to carry out their work for the rest of their life.

However, our formal educational methodologies seem to have advanced less rapidly than many other aspects of society. Our formal education systems seem out of step with rapid-growth hi-tech industries and online communities, and more akin to those old-fashioned, fusty domains of industry that haven't yet caught up with the times.

*Informal* means of education are accelerating dramatically, due to largely to computers and the Internet. Software primarily labeled as "educational" has made a relatively small mark on the world, yet Wikipedia, YouTube, Google, ArXiv, CiteSeer, and other such resources have had a tremendous impact and are doing a remarkable amount to spread knowledge throughout the population of humans with regular Internet access (along with spreading entertainment, nonsense, and a lot of other things).

At the high end, MIT and other universities are putting more and more of their curriculum online. For example: anyone with a computer, an Internet connection and a reasonable high school background can get a thorough education in computer science and software engineering via viewing free online lectures, reading free online textbooks, asking questions on free online forums, and practicing programming using free compilers and development environments, etc.

Internet technology provides amazing and accelerating means to bring people together to allow them to teach each other. Online forums are one example ... another is language learning websites, that allow, say, a Japanese speaker wanting to learn English to connect with an English speaker wanting to learn Japanese, for mutually educational multilingual education.

And the interactivity of many of these knowledge resources is important. A child researching a school project using Wikipedia may notice an error or omission in Wikipedia and update the site accordingly. A child can study animation and then upload their animation to YouTube for others to comment on. The boundary between learning and doing breaks down.

All this is well known. What is not sufficiently discussed is where this trend is leading us.

## Education Obsoletes Itself

The schools of the future are going to look nothing like the schools of today. If indeed there are schools at all.

Even given all the educational affordances provided by modern technology, there may still be a value for schools of some sort, for social reasons. But if they do exist, schools of the future will serve more to regulate students' educational interactions with the world at large, rather than to disseminate information directly. Students will learn by doing, and learn by exploring the Net and interacting with people and artificial agents from around the world ... and teachers will be there to gently nudge and guide this activity.

The main reason education isn't this way right now is inertia. And this inertia is very strong, especially in places like the Orient where education is based almost entirely on rote, with minimal emphasis given to initiative or creativity.

But this aspect of society will change — because it has to change ... because old-fashioned schools are getting less and less useful at preparing people for newfangled society.

Where this leads is to the end of the distinction between education and plain old everyday life. If you learn by doing, and you need to constantly learn while doing anything due to the constant influx of new information ... then where lies the distinction between learning and living?

This is plain vanilla (or maybe rainbow-colored?), hippy-dippy "progressive education", really — but what's not sufficiently appreciated is that it's going to happen, not because it's a nice and friendly and creativity-encouraging way to do things, but because it's going to be judged necessary for preparing students to deal with a rapidly-changing and increasingly information-rich world.

And that's without even mentioning the really groovy stuff — the possibilities for education afforded by, say, cranial jacks feeding knowledge directly into the brain ... or virtual worlds allowing students to try out new experiences in a manner partially self-guided and partially remote-controlled by others ... etc. etc. etc.

These various advanced educational technologies could potentially be shoehorned into the old-fashioned, rote-based, one-size-tries-to-fit-all, learning-separate-from-doing style of education ... but doing so would plainly squander most of their potential.

Education wants to be free ... and free of schools and traditional educational methodology ...

## Moral of the Story?

And so as technology advances and society adapts, "education" will disappear as a separate category and pervade through life ... for those who want to keep growing.

On the other hand, some folks may wind up choosing to spend their time ignorantly pursuing repetitive pleasures in simulated worlds, or other similar activities. But these folks won't need schools nor much education either. So one way or another, education per se will soon be a thing of the past.

The moral: promote informal learning that pervades life ... it's the way of the future. Use formal learning set apart from life as a tool when it's the most valuable choice given our current situation, but be aware that it's decreasingly relevant as the future unfolds.

# Happy Goldfish Bowl to You, to Me, to Everyone

Isaac Asimov's classic 1956 story "The Dead Past" describes a technology that lets everyone spy on everyone else everywhere. The government tries to keep it secret but some rebels reveal it. A government official (Thaddeus Araman), once he realizes the technology has been made irrevocably public, utters the following lines to the scientists who released it:

*"Happy goldfish bowl to you, to me, to everyone, and may each of you fry in hell forever. Arrest rescinded."*

What happens if, one day, advanced technology allows everyone to see everybody else's activities? This technology has been labeled "sousveillance" — all seeing all — as contrasted to surveillance, in which the few observe the rest. David Brin's book *The Transparent Society* treats the theme in wonderful detail.

Thaddeus Araman considered sousveillance a bad thing — and many would agree.

But on the other hand, in Asimov's story, the government was just leaving the "sousveillance" technology dormant — it wasn't using it for its own purposes, to implement Big Brother or some yet more dire alternative.

So if sousveillance sounds scary to you, consider: What if the alternative is only the Chosen Few getting to look into everybody else's minds? At least sousveillance lets the watched watch the watchers.

And in addition to its potential for avoiding *sur*veillance, sousveillance could have other benefits as well.

There's no doubt that sousveillance would wreak havok with current notions of sociology and psychology. Self-models would never be the same! But self-models in their current versions cause a lot of problems — a bit less self-delusion, induced by sousveillant transparency, could potentially leave us more relaxed, realistic and cooperative.

One challenge in a sousveillant society would be maintaining diversity — resisting the pressure for conformity that would come from having one's deviances made public.

On the other hand, if everyone's secret freakishness is revealed, perhaps this would make everyone more tolerant of everyone else's freakishness!

And what about "strong sousveillance" — where everybody can see into everybody else's minds?

Certainly, strong sousveillance would open up tremendous possibilities for novel forms of creative consciousness and cognition. We could think and feel together in ways that aren't possible now.

And yet the difficulty of *understanding* others' minds should not be underestimated. In a sousveillant society, not just intelligence but *incomprehensible intelligence* will be at a premium.

Would strong sousveillance inevitably lead to the formation of a unified overmind, or at least a mindplex with emergent social-level coherent self and reflective awareness, along with individual-level self and reflective awareness?

At this stage it's hard to say whether (either weak or strong) sousveillance would prove a good or bad thing in term of the principles of Cosmism ... it may depend upon the details.

But it does seem clear that the potential for various forms of sousveillance would be a positive — if it were managed in a spirit of growth, joy and choice. Experimentation with sousveillance could be fascinating, and lead to all sorts of social and mindplexish patterns we can now barely imagine....

## Meditation & Beyond

Many people have reported achieving various sorts of profound insights through meditation practice.

Some say it has allowed them to escape human reality altogether, and enter into other dimensions of being, or to become enlightened and escape the very realm of "being".

I think this is great ... but it's also interesting to look at meditation from a physical-reality-centered perspective. In this view, meditation is a mind-state associated with a certain human-brain-state. It may have certain profound advantages relative to the "ordinary waking consciousness" and other mind-states that humans more commonly occupy. But nonetheless it is tied, in a way, to the human brain. The book *Zen and the Brain* and others explore this connection in depth.

My suspicion is that, as human brains are enhanced and expanded, yet more amazing and insightful forms of experience will be found.

Meditation is largely about having the flow of thoughts, feelings and habits stop. It takes most humans great practice to achieve this. On the other hand, appropriately designed cognitive systems could potentially achieve this instantaneously, merely by "flicking an internal switch."

It's interesting to think that AI's — or neuromodified humans — could switch back and forth between meditation and practical, highly task-focused consciousness "at will." Or they could have one portion of their brain meditate while the other portion carries out various other activities ... etc.

## Enlightenment

*Students achieving 'oneness'*

*will move on to 'twoness.'*

*— Woody Allen*

*Enlightenment,*

*don't know what it is*

*— Van Morrison*

Some individuals have professed to reach a state of "enlightenment", in which they go beyond human cares and limitations, become one with the universe, and experience an elemental perfection that literally cannot be imagined.

I am sure this is an amazing, rewarding state of mind — but one has to recognize that it also has much in common with addictive states of minds achieved through drugs, romantic love or other means.

Many states of mind have the property that, when you're in them, you feel they're the most important, wonderful thing in the world.

Then once you get out of them, later, you wonder how you ever felt they were quite that essential.

From a neurological perspective (which I stress is not the only valuable one!), enlightenment may be a bit like these — except it's a more powerful "attractor", and once you're in, you don't get out.

I wouldn't want to trivialize the amazing experience that some individuals call "enlightenment." Yet I'd hesitate to classify any state of consciousness as absolute *perfection* even if, in some sense, it *tells* you it is.

Of course not all "enlightened masters" do classify their states as perfection — not surprisingly, the rhetoric surrounding "states of mind beyond description" becomes subtle, ironic and paradoxical.

And of course, from the point of view of First, putting any experience in a labeled box (like "enlightened" or "perfect") is irrelevant — labels are Third anyway. Enlightenment is about experiencing First as First and not mixing up its pure Firstness with the tangle of relationship. When you're meditating and you start to ponder or perceive some relationship and enter the realm of Third, the Firstness of that Third becomes apparent — and you're back to the Firstness of First again. (And of course, this paragraph, which indulges in the dubious amusement of relating enlightenment to Peircean categories, is just another Third, which —)

One wonders if yet deeper and more amazing "enlightened states" could emerge from minds associated with more powerful cognitive architectures than the human brain.

## Beyond Enlightenment?

Profound as enlightenment is from its own perspective — it seems to have its limitations, from a pragmatic view.

It's well worth noting that in human history, enlightenment seems to be anticorrelated with some other valuable things, such as deep scientific, mathematical or engineering creation.

What might the reason for this be?

Quite possibly, maintaining mental purity is difficult for our feeble human minds, so that it consumes most of our resources, not leaving much for anything else.

Might more powerful minds than humans be able to maintain the exaltation of enlightenment while also effectively pursuing activities requiring deep analytical thinking?

Could improvements to our cognitive architecture obsolete the dilemma between enlightenment and creative productivity?

## Psychedelics

Like meditation, psychedelics are a tool that many humans have used to achieve profound and insightful states of consciousness going far beyond the ordinary ones we experience in our daily lives.

Compared to meditation they have different pluses and minuses. They can have profound transformative effect, and take you into amazing places, without requiring years of preliminary practice first. On the other hand — like meditation but perhaps even more profoundly so — they can induce all sorts of confusion, states in which one part of the mind fools the other and so forth.

The criminalization of psychedelics in most modern societies has made it difficult for their potential to yield positive transformation in the contemporary context to be thoroughly explored. This is a real shame.

However, it seems likely that once technology has advanced further, we will have far less crude means of altering our state of consciousness. Psychedelics and meditation will then be viewed as blunt instruments for modifying brain dynamics. Once brain dynamics are better understood ... and especially, once we have either nanotech for pliably modifying the brain, or uploading that puts our minds into more pliable substrates ... then we'll be able to shape our state of mind the way a master sculptor shapes a hunk of clay.

At that stage, "drugs are bad" won't be a moral judgment, but rather a realistic assessment of their relative effectiveness!

# Might There Be Intelligences in Other "Dimensions"?

Many people, after having certain meditative experiences or taking certain psychedelic substances (especially DMT), emerge with a strong intuitive sense that they have been communicating with intelligent transhuman beings in some other "dimension" — a dimension quite close-by to us, but normally inaccessible to us due to the nature of our mind-architecture and self-structure....

Some folks, such as Terrence McKenna, have hypothesized that the technological Singularity will put us in touch with these beings (which he whimsically labeled "nine-dimensional machine-elves"!), via allowing us to occupy more flexible mind-architectures and lose the restrictions of the human self...

Interestingly, this hypothesis that we'll contact such beings after the Singularity *is* verifiable/falsifiable... we just need to create the Singularity to find out!

If nothing else, this line of thinking serves to remind us that it's mighty hard to meaningfully chart what might happen after Singularity. After all, if McKenna is right and post-Singularity we will contact these beings and ingest information from them or in some sense join their world — then from that point on the direction of our mind-evolution will be quite independent of any detailed prognostications we might make now...

These ideas seem related to Philip K. Dick's experiences on February 3 1974, which are nicely recounted in the biography *Divine Invasions* — and during which he says he received medical information from alien minds, which he would have had no way to find out through ordinary means, and which he later claimed to prove valid via conventional medical examination. (Of course, though, this instance of the mysterious transmission of medical info to Dick — assuming it really happened — could be explained via simpler psi phenomena, not requiring the postulate of alien minds!)

This is weird stuff from a contemporary-Western-culture perspective, and *may* be best understood as nothing more than strange experiences generated by human brains under the influence of various (ingested or self-generated) chemicals.

However, the Cosmist perspective urges open-mindedness. The universe is a big place — perhaps in senses beyond the ones modern physics acknowledges —

and we likely experience only small fragments of it. That certain states of mind could allow some humans to experience chunks of universe inaccessible to ordinary waking human consciousness, is certainly not impossible.

# How Real Is Reality Anyway?

After all, the empirical world of electrons and baseballs and such is known to each of us only via inference and extrapolation based on our (lifetime of) sense data.

That is: the "empirical world" itself is, from a subjective perspective, something each of us invents for ourselves, elaborating on patterns we recognize in our sense-data (including linguistic communications from others).

So the question is which of our sense data do we choose to trust — i.e. do we mistrust the data received while under the influence of DMT while accepting the data received during ordinary waking consciousness ... or do we take a more open-minded view?

I'm not saying people should ascribe a profound reality to their every passing delusion, hallucination, etc.

Just that the distinction between reality and invention is not that clear — so we need to be careful about dismissing something just because it diverges from the sociopsychological construct we think of as "empirical reality."

What is the difference between a "reality" and a "collective invention that evolves dynamically and creates new forms that it feeds back into the minds of the inventors"?

## The Puzzling Nature of "Simplicity"

This odd issue of DMT aliens sheds an interesting light on the nature of the "simplicity" that underlies the Occam's Razor heuristic.

To nearly everyone who hasn't communicated with these beings themselves, the hypothesis "it's a drug-induced delusion" seems the simplest explanation for the phenomenon at hand.

Yet, to nearly everyone who has communicated with the beings, the hypothesis "they're real, autonomous, nonhuman beings of some sort" seems the simplest explanation — because the sense of independence and alien-ness and intelligence these beings project is so powerful, it just seems intuitively absurd that they could be produced by the mere human brain.

A mind's assessment of simplicity is not independent of its experience base! So, the patterns a mind sees, being dependent on what a mind has experienced, are a function of the mind's own beliefs, ideas and memories.

## Beyond Insanity

I've been called "crazy" more than a few times in my life, due to having eccentric ideas and a nonconformist personality. Thankfully, though, I've never shown any signs of "mental illness."

But the phenomenon has always perplexed me ... I've often wondered on questions like

*What is "madness"?*

*Why have there been some notably insane geniuses?*

*What can lunacy teach us about mind and reality?*

Exploration of these questions is made difficult by the generally convoluted way the notion of "insanity" is defined and conceptualized. The situation is somewhat similar to the one with "free will" — we all sort-of know what the term means, even though it's wracked with conceptual conflicts. And just as I find "natural autonomy" a more useful concept than classical "free will", here I will describe a notion of "mental disharmony" that I find clearer and more useful than "insanity."

I will use a lot of quotation marks in this chapter —

when I want to emphasize that a term like "insanity" is being used in a manner that accords with communicative convention but that I might find hard to defend in a truly rigorous way.

## Madness, Ridiculous and Sublime

In my teens I tended to romanticize insanity somewhat, due to reading about mad geniuses and the "fine line between genius and insanity" — but as my later life brought me into contact with more mentally unhealthy people, I learned that by and large "madness" brings a lot of pain and tedium, and is not that exciting after all. For example, I once had a paranoid schizophrenic in-law make serious threats to murder my 4 year old son due to his alleged role in the international political conspiracy against her. Not so romantic.

I exchanged many emails last year with a bright, enthusiastic young Canadian who uses the online handle *flamoot*, who sent a host of messages to futurist email lists, regarding an implant that he believes someone (aliens, or the NSA, or the Scientologists, etc. — the story varies) has planted in his brain. He hears the implant speaking to him and giving him instructions and comments regularly, and understandably becomes quite frustrated with the intrusion into his mental space. He thought there was a metal implant in his head, and had his head carefully scanned in a neuro lab, but they didn't find anything.

I also recall the former girlfriend of a Finnish friend, who believed that participants in a conspiracy against her were modifying all the songs in her MP3 player, causing them to have strange lyrics containing secret messages about her. No matter how much one argues with her that this is highly technically implausible (and she has the technical training to understand such things), she maintains the belief with absolute confidence. She just will not accept the notion that her own brain could be misinterpreting the lyrics somehow — that her cognition could be driving her perception.

From the "consensus reality" perspective, these particular instances of "mental illness" are just ridiculous and upsetting to the individuals involved, and seem to have no positive consequences. But there are many other cases of apparent insanity playing a role in wonderful creative inspiration.

William Blake literally saw angels sitting on the trees near his house, as a child. No wonder he wrote such lovely poems about them! He treated them as metaphors in profound ways; but to him they were more than metaphors as well.

Van Gogh made his most beautiful, amazing paintings while in the midst of an emotional breakdown that at one point caused him to sever part of his ear and mail it to a girlfriend.

Further historical examples abound.

Reading Blake, I can't help feeling that "madness" gave him access to certain aspects of fundamental reality that are shut off to most humans, due to the constraints placed on our minds by our conventional cultural perspective, our "sanity."

But if Blake's "madness" gave him fundamental insights, what about flamoot's?

In his intriguing memoir of his life among the Ba-Banzelle pygmies of central Africa, American Louis Sarno notes how when the pygmies sing and dance and play music, the forest spirits come out to play as well. The leaves on the forest floor leap up and dance in strange forms, in sync with the music. The pygmies take this for granted, and when Sarno is with them, he does as well. To consensus modern civilized reality this sounds like pure madness; but to the pygmies, *not* seeing the forest spirits would be an unfortunate personal mental deficiency.

There's no doubt that certain kinds of mind-state commonly classified as "mental illness" involve getting beyond the everyday perception of self and reality, in a way that doesn't happen during most peoples' ordinary lives. This additional open-ness can stimulate creativity – like Blake's poems, Van Gogh's paintings, and the music and dance of the pygmies. And it can do this in highly nontrivial ways.

"Madness" doesn't work its sometime creative magic merely by throwing exciting new forms into the mind – it's not a matter of someone hallucinating wild movies and then writing them down. Rather, when "madness" is entwined with creation, it develops in a profound way, via exerting a transformative effect on the self and its relationship to reality. The forest spirits embody the oneness between the pygmy and the forest, the aliveness that the pygmy feels in common with the forest. Blake did not just see the angels, he felt them in his heart; he accepted them as his friends and part of his extended self. The world Van Gogh portrayed in his paintings was something he perceived as a fundamental reality.

A fairly high percentage of creative "lunatics" have had bipolar disorder (formerly called "manic depression"). I have been close personally to people with bipolar disorder, and my impression is that, as a side-effect of their ongoing oscillation between profound depression and extreme elation, *bipolar people are continually faced with the task of rebuilding their selves.*

In an intense depressive episode the self-model basically collapses and becomes unworkable and unsustainable and panic or terror result; then as it passes and the manic phase approaches, the self-model is rebuilt again, and you once again have a "whole person." But each time the self is rebuilt, there's the potential for something new and interesting to arise. This is very different than a typical calm, stable person who maintains the same steady sense of self for their whole life.

This is not to say that calm people cannot grow and adapt and expand their minds — of course they can. And bipolar people can be extraordinarily repetitive, if they repeat the same depressed complaints and the same manic delusions of grandeur each time they go around the cycle. But bipolar disorder, like some other forms of "mental illness", does contain a sort of built-in mechanism for innovation — and at the fundamental level of the whole self, not just at the level of individual thoughts or ideas.

But "madness" opens up the mind in a much less controlled way than meditation or related practices, and even in a less controlled way than psychedelic drug use. A consequence of this lack of control is an increased social difficulty — when you meditate or take drugs, you may go "way out there", but you have the chance to do it at a time and place of your choosing. When you "go crazy" it's a bit different: an episode of highly nonstandard perception and cognition may come upon you with less warning, in a context where it's more troublesome.

# Mental Disharmony

Stepping away from the fascinating particulars, what is the general crux of "insanity"?

Is it just "thinking differently" than the consensus? If so then insanity becomes a dramatically culturally relative concept. A typical modern person has beliefs and habits that would have been considered insane in ancient Egypt or ancient China, and would be considered insane *today* in the isolated Stone Age tribes in the Amazon jungle.

My suggestion is that, instead of "insanity," it's often more useful to think about "mental disharmony."

We can ask, about a certain mind:

- Does the mind work together with the other minds around it, to form a productive, coherent cognitive system, manifesting collective, joy, growth and choice?

- Do the different parts of the mind work together with each other harmoniously, forming a coherent overall system, not wracked with conflicts?

Of course perfect harmony is hard to come by — but in many cases of diagnosed "insanity" there are *huge* disharmonies of one or both of these kinds. So many "mentally ill" people are wracked by inner conflict of one kind or another.

A modern American thrust back into ancient Egypt would initially pair internal harmony with external disharmony — but before long the situation might start to drive him nuts and he might go insane internally as well. Either that or he would need to "go Egyptian" inside as well as out.

And the contemporary Muslim suicide bomber may be sane in his own culture though insane in American culture.

There are many, many ways for individual and extended minds to be disharmonious — the modern psychologist's DSM IV manual enumerates many that are common in current Western culture, labeling them forms of "mental illness." Some of these categories (like bipolar disorder) seem to be innate to the human brain; others are more culturally dependent, e.g. the disease of "hysteria" which was common among Victorian women but seems to have vanished along with the female sexual and social oppression of that era.

## Some Lessons from Lunacy

Thinking about "insanity" in terms of mental disharmony makes its plusses and minuses easier to comprehend.

Harmony is wonderful — but it can also be a dead end, a local optimum. If everything is working happily and smoothly together, there may be no motive for change, even changes for the better. Sometimes dislocating one's internal or extended mind can be a good thing, and can lead to the emergence of novel structures, dynamics and possibilities. Getting outside the normal ways of thinking, perceiving, acting, being and self-modeling can be beneficial — even if it involves some confusion and conflict.

Too much disharmony leads to miserable chaos. Too little, runs the risk of falling into a rut.

A little "madness" can spur the mind in amazing new directions. Too much can leave it on the floor gibbering and flailing in the throes of its own contradictions.

All of which is a special case of the obvious and general rule that: in finite systems, there are sometimes subtle trade-offs between Joy and Growth.

And what about genius and madness?

Being "insane" certainly won't make you a genius — but it can do two things

- By bringing you away from the consensus, "madness" can give you new ideas, which can then be used in a genius way if other preconditions are met (e.g. the right sort of attitude, some innate talent, and large masses of appropriately structured hard work)

- By bringing you away from the consensus motivational system of your society, it can free you up to adopt novel motivations, which may sometimes be ones that cause you to devote large amounts of energy to creative pursuits (and passion and obsession are clearly correlated with great achievement and "genius")

Most certainly: if you don't have a tendency toward "mental illness", please don't try to induce it!  There are other ways to be fantastically creative, with fewer destructive side-effects!

But do consider how you might introduce wild new ideas into your mind, and how you might reshape your motivational system away from your culture's consensus in the interest of greater joy, growth and choice.

And if you do have a tendency toward "mental illness" — by all means, milk it! Don't let it destroy your life, but also don't ignore the gifts it can convey. Arrange your life so that whatever your "madness" is has minimum destructive impact, and make the most of the innovations in concept and self-model that your particular brain chemistry and mind-organization brings you. But also be aware that to turn these innovations into anything dramatically wonderful — in the domain of creative works, or personal growth — is bound to require massive amounts of hard work ... and in most cases some rationality and self-discipline mixed in with the creative madness.

# The Future of Madness

Once science advances a bit further, there won't be any more "insanity."

Good riddance!

But there *will* still be the possibility to jiggle your mind a bit (or even more than a bit) — to perceive and/or analyze things differently than everyone else, to break down the self-model and rebuilt it from scratch.

The sorts of insights that a few geniuses have acquired from their painful insanity, will be achievable in other ways, without all the unpleasant, soul-scarring side-effects.

The dilemma of madness versus mental conformity will be obsoleted, in favor of

*Cosmism beyond sanity and insanity.*

## Cetacean Minds

Strange as it may sound, it is quite possible we share our planet right now with other creatures equal or even greater in general intelligence to us.

I'm not talking about invisible aliens or anything bizarre like that (though, hey, those might be around too, you never know!) — I'm talking about cetaceans. Dolphins, whales, and so forth.

They have huge associational cortices ... i.e. huge parts of their brains not devoted to anything obvious like perception or motion. And they've proved capable of learning all sorts of complex things, including language with serious syntax. We know they communicate among each other in sophisticated ways, for instance mentioning each other by name during social conversations; but we don't really understand much of their language or to what extent it really is a "language" in the human sense.

Maybe their big brains are largely just huge mapping systems for getting around in the ocean. But there's much evidence that cetaceans have extremely complex social structures, just as humans do — and it's widely conjectured that human intelligence largely arose out of the need to navigate human society (so that social complexity and cognitive complexity recursively "pumped each other up").

It is tempting to hypothesize that cetacean consciousness, compared to human consciousness, has more to do with *shaping and flowing* and less to do with *causing and building.*

Ultimately, though, we don't really know how smart cetaceans are, or in what ways they're smart, because we've put precious little resources into serious studies of cetacean cognition and communication. We may have an intelligent "alien species" right under our noses ... but rather than studying them with maximal intensity we're murdering them as part of our commercial fishing practices!

What cetacea haven't done, that we humans have, is build complex tools. They do (like some birds and apes) use tools, but they don't make tools to make tools, and so forth. This is why we are far more likely than them to launch a Singularity — even if they are "smarter" in important senses.

But still, we may have something to learn from them —
and advanced technologies may eventually break down
the barriers and allow us to communicate with them. When
an AGI serves as translator between human and dolphin

—

## Questing Extraterrestrial Intelligence

What about the search for intelligent lifeforms elsewhere in our physical universe?

This is certainly a very worthwhile pursuit: the principle of Growth implies being open to all sorts of possibilities ... and the existence of intelligent lifeforms in other star systems, galaxies etc. is certainly a feasible possibility.

The Singularity notion casts SETI in an interesting light. Suppose that, shortly after developing advanced technology, a civilization tends to reach a Singularity and its members then advance their intelligence dramatically. In that case, it might well be that there are very few civilizations in the universe at the stage of advancement where they would be interested to communicate with us.

John Smart has suggested that once intelligences get smart enough, they want to maximize their intelligence via packing their processing units into the smallest possible volume of space (to minimize communication delays that are inevitably proportional to the distance betwen processing units, according to the physics of special relativity). So, he suggests, the smartest minds will become so dense as to become black holes. Maybe the black holes we detect with our radio telescopes are actually superhuman minds.

On the other hand, perhaps superhuman minds have somehow distributed themselves throughout the universe, expressing their thought-processes in terms of the quantum or subquantum vibrations and interactions of particles. In that case there would be alien superhuman mind within us and everything we interact with. Perhaps what we think of as "quantum noise" is actually highly complexly patterned — but they're patterns we're unable to detect or understand, because they're the thoughts of vastly more advanced intelligences.

Stanislaw Lem's novel *Solaris* remains the most insightful thing yet written about SETI. It features an alien ocean that is clearly intelligent in some sense – probably superhumanly so — yet is so very alien in its mind-patterns that we humans have no hope of communicating with it in any useful way.

Greg Egan's novel *Diaspora* presents a possible solution to the problem Lem poses: build a series of minds forming a bridge between ourselves and the alien. In the Solaris case, this would involve building

- a mind halfway between us and the ocean
- two additional minds: one halfway between us and the new "halfway" mind, one halfway between the ocean and the new "halfway" mind
- four additional minds ... etc.

Implementing this plan in any particular case may involve some "minor engineering difficulties" — but it's a fascinating approach.

Then there is the sobering possibility that we may have alien human-level intelligences right here on Earth — Cetacea — yet are unable to communicate with them beyond a trivial level (so far) due to the profound cognitive differences.

Let us search for extraterrestrial intelligences, by all means! But let us not be too anthropocentric in our ideas about what sorts of intelligence to search for. Pattern space is a very big place; minds vaguely similar to humans are most probably a tiny subset of the space of all minds feasible within our physical universe. And the physical universe as we now understand it may be nothing more than a tiny subset of the actual universe.

# The Mystery of Psi

Psi powers — ESP, precognition, psychokinesis and the like — is one of those topics that seem to polarize human opinions (I'm referring to the modern era, not to prior periods when their reality was taken for granted by essentially everyone).

Some folks, largely on the strength of their own personal experiences, take their existence as an obvious truth.

Others reject them without much consideration — because current scientific theories fail to indicate any clear mechanism for their operation, and because so many charlatans have falsely claimed various psychic powers and then been debunked.

I encourage everyone to read Damien Broderick's fantastic book "Outside the Gates of Science," which carefully reviews some of the scientific evidence that psi exists, and also analyzes its apparent limitations. Before reading his book and exploring the scientific literature it cites, I was basically 50-50 regarding the reality of psi; but after reading it I've shifted to 90-10 in favor.

I strongly suspect that during the next half-century a scientific model of psi will be created, which will tell us something about how psi works and under what conditions it has what capabilities.

The implications of psi for future technology and Cosmism in general are not particularly clear at the moment, because psi is so poorly understood.

But at a broad level one message comes through clearly from the data Broderick reviews: the universe is more tightly and intricately interconnected than the modern Western world-view admits. Patterns in parts of the world that appear to be separated and uncorrelated, are actually tightly tied together. We don't know much about how this works but the implications are potentially quite broad ... certainly this meme resonates with the notion of the "mind of the universe" that was mentioned earlier in this Manifesto. Psi could potentially reflect processes, as yet largely opaque to humans, by which the universal mind carries out its coordinated intelligent dynamics.

# Immortality: Should We Want It? What Does It Really Mean?

Some people want badly to live forever.

This has long been the case – an amusing tidbit of my family history is that in the early 1960s my father led an organization called the Student League for the Abolition of Mortality (SLAM). And of course, the Taoist sages sought immortality long before my dad, via all sorts of practices that seem very strange to us today; and they were preceded by tribal shamen.

But until recently immortality was an unrealistic goal — unless interpreted very broadly, as in living forever through one's children, one's artworks, one's connectedness with the world, etc.

These "looser" ways of living forever are deeply meaningful — they do, in a strong sense, involve one's mind living on after one's body's death. Each mind is a set of patterns, and in these ways many of the core patterns of one's mind can live on independently of the body that previously carried them.

But one thing that these looser forms of immortality don't give you is immortality of the *self.*

Advanced technology, however, may soon make this stronger form of immortality possible — maybe through pharmacology or other biological means like stem cell infusions, maybe through nanotech bio-repair bots, or maybe via more controversial mechanisms like uploading.

However the technological details work out, for many of us, this prospect is tremendously exciting. For Cosmists who want to keep growing, choosing and enjoying, the prospect of not dying is often an appealing one.

## Is Immortality Desirable?

But, although some people want badly to live forever, others badly want *not* to: they consider death intrinsic to the meaning of their lives.

Neither of these desires is intrinsically unhealthy or foolish: they just reflect different self-models, different ways of interpreting the relationship between the self and the world, and the relationships between selves that exist at different times.

Imposing immortality on those who don't want it, is nearly as bad as imposing death on those who don't want it.

Of course, either death-avoidance or death-seeking can be carried to unhealthy extremes.

Obsession with danger sports or dangerous drugs or outright suicidalness — i.e. the Freudian "death-wish" — are among the obvious examples of unhealthy death-seeking.

Unhealthy death-avoidance could manifest itself as a reluctance to do anything even slightly dangerous, for fear of sacrificing one's potentially infinite future life — and I have actually met some radical futurists who suffer from this sort of issue!

As Ben Franklin said: "Moderation in all things, moderation included."

## Would Immortality of the Individual Be Good or Bad for Society?

It's not clear what mix of death and immortality (or extreme longevity) is optimal on a society-wide basis, in terms of maximizing joy and growth.

Clearly the death of brilliant, productive individuals with brains full of knowledge and experience is a waste.

On the other hand, there's some truth behind the *bon mot* that "Science advances one funeral at a time" — old minds can get hide-bound with habit; new perspectives can lead to accelerated advancement.

It seems rather clear, though, that whatever the optimal balance is, currently human existence is tipped way too far toward the *death* side of things ... due to brute biological necessity.

Helping with life extension research is one of the most important things any person can do today

## Continuity of Self

Perhaps the deepest issue regarding immortality is "continuity of self."

If I live a billion years, and change by 1% each year, then before long I may become something that has no resemblance or commonality at all with what I am today. In what sense will that be "me"?

One way to think about this is: If a mind is changing smoothly and incrementally (maybe quickly, maybe slowly), *and* at each stage the phenomenal self of that mind *feels like* it's preserving itself as it changes — then one has meaningful mental continuity. Perceived continuity of self, combined with empirical continuity of mind-stuff, is continuity enough.

This kind of "continuity of self" probably places limits on how fast a mind can evolve ... if you evolve too fast, the self can't feel itself evolving, and the subjective experience will basically be that of one mind dying and another getting born in the same vehicle.

One can imagine a future in which a certain group of minds abandons continuity and evolves super-fast into something amazingly advanced, whereas another group maintains continuity and as a result transcends more slowly.

## Continuity and Growth

Continuity emerges in Cosmism as part of the value of Growth. Growth is not just about new patterns forming as time unfolds — it's about old patterns growing into new and richer ones.

Without continuity of self, selves don't grow, they vanish and are replaced!

On the other hand, if continuity of self slows down the evolution of mind, then it may be a bad thing to the extent that it restrains the growth of minds.

The Choice principle suggests that each mind be allowed to judge this tradeoff for itself, inasmuch as this is possible.

One hopes there are ways of obsoleting this dilemma, which are not yet apparent!

## Uploading

One of the more exciting and controversial approaches to immortality is "uploading" — the transfer of the human mind from the human brain/body into a different substrate, such as a digital computer system or a robot.

As my favorite uploading joke goes: " Uploading – it' s a no-brainer!"

We don't know exactly how to do this yet, but it seems likely we'll figure it out during the next few decades ... at worst during this century.

Even if it turns out that the brain/body's functions can't be emulated effectively by a digital computer, uploading should still be perfectly possible — one would just need to create the right sort of computer, such as a quantum computer or a quantum-gravity computer, or whatever.

Yes, it's *conceivable* that physics somehow prohibits us from copying the details of a mind out of a brain so it can be projected into some other substrate — but based on all we know about science, that seems extremely unlikely.

The philosophical questions raised by uploading have been debated *ad nauseum* within the futurist community, to the extent that discussion of the issue is forbidden on some futurist email lists (due to the repetitive nature of the discussions).

The Cosmist perspective is simple: based on everything we know today, it seems extremely likely that the uploaded version of Bob Jones is "the same mind as" Bob Jones, in the same sense that Bob Jones at 4:15 PM on a certain day is "the same mind" as Bob Jones at 4:14 PM on that day.

The patterns comprising Bob Jones's mind in its human-brain/body version, would also be there in its uploaded, computer-substrate version. So from the point of view of Third, it's all the same Bob Jones.

And, from the point of view of First, it's *all* one big, small moment of experience.

Worries about continuity of self and mind would seem to be assuaged by uploading scenarios in which the biological brain is transferred into a digital simulation one neuron at a time. And then, perhaps the digitally simulated brain is gradually, slowly transformed into something less and less resemblant of a legacy human brain.

# What About Weaker Forms of Uploading?

More controversial are weaker forms of uploading, such as creating a computer-based mind that emulates Bob Jones based on information gathered from things he wrote, videos of him moving around and speaking, and so forth.

One question is whether such weak-uploads could be created with sufficient fidelity to capture the "essence" of the individual's mind and self. My guess is yes, but we don't know enough to say for sure.

Another question is about continuity of self: when Bob's weak-upload first gets up and running, it won't necessarily feel like a continuation of Bob's previous thought-stream. The "continuity" aspect of growth exists only weakly here ... this is less a growth of existing patterns, than a replacement of prior patterns with new, highly similar ones. Consequently this seems to be an inferior form of uploading — but still, quite an interesting one.

# Does Human Mind Make Sense Without Human Body?

The mind is tied to the body, more closely than some (especially modern computer geeks) want to recognize. Mind emerges from body as well as brain.

This raises the question of whether some of the commonly envisioned forms of uploading even make sense, experientially. Perhaps putting a human mind in a PC is intrinsically senseless, because that mind would very rapidly drift into a wholly nonhuman state of being due to its radically nonhuman embodiment.

In this case, the key to making uploading work psychologically would be to upload the mind into a receptacle that shared the key experiential features of the human body, but lacked its more problematic issues — like its rapid rate of decay!

One nice thing about uploading is that it can potentially be tried over and over, with the same mind being placed into a vast number of different substrates with different properties. In this way a single stream of awareness could diverge into several different streams emerging from different substrates. And why not?

# The Prospect of Transhuman Artificial Intelligence

Of all the amazing technologies on the brink of being created, one has implications far beyond any others: the creation of superhuman AI. This is the reason that for years I have spent most of my own working hours on the development of artificial intelligence theory and technology.

But the term "AI" has become a bit diffuse lately – it pays to refine it a bit.

"Narrow AI" systems with task-specific intelligent capabilities but without human-level breadth of intelligence may be very useful — for instance at liberating humans from toil and creating new avenues for us to enjoy ourselves and do science and art and so forth.

But all this pales before the implications of creating non-human-like "artificial general intelligences" (AGIs), which have an ability equal to or greater than that of humans to transfer knowledge from one domain to another, to create new ideas, to enter a new situation, get oriented, and understand what are the problems they want to solve.

Current AI programs are very far from possessing general intelligence at the human level, let alone beyond — but an increasing minority of AI researchers agree with me that superhuman AGI may come within the next few decades ... conceivably even the next decade, and almost certainly during this century.

We don't yet know what the quickest, best path to powerful AGI will be. A number of approaches are out there, for instance:

- emulating the human brain, at some level of abstraction
- leveraging knowledge resources like Google to create systems that learn from patterns in texts
- developmental robotics, that begins with an unstructured learning system and gains experience via engaging with the world
- evolving artificial life forms in artificial ecosystems, and nudging them to evolve intelligence

My own bet as an AGI researcher is on an *integrative* and *developmental* algorithmic approach. What I'm working on is creating a software system that combines different AI learning algorithms associated with different kinds of memory, and using it to control virtual humanoid agents in online virtual worlds, as well as physical robots. The idea is then to teach the young AI in a manner similar to how one teaches a human child, interactively leading it through the same stages of development that young humans go through. Except the AI won't stop when it reaches the level of adult humans — it will keep on developing.

The hard part, in this sort of approach, is getting all the different AI algorithms to interact with each other productively — so that they boost rather than hamper each others' intelligence. Evolution created this kind of cognitive synergy in the brain — in building artificial minds, unless one tries to closely emulate the brain (which brain science doesn't yet tell us enough to do), one has to specifically engineer it. It's not easy. But sometime in this century — maybe sooner rather than later — somebody is going to get to advanced AGI, whether via an integrative algorithmic approach or one of the other avenues.

# Far Beyond Humanity

What will this mean for us humans, the advent of superhuman AGI?

We don't know and we can't know. Anthropomorphizing nonhuman minds is a profound danger, when thinking about such things.

Of course, a mind that we create can be expected to be far more humanly comprehensible than a "random intelligent mind" would. In the period shortly after their creation, we will likely understand our AI offspring quite well. If all goes well, they will cooperate with us to solve the various niggling problems that plague our existence — little things like death and disease, and material scarcity.

But we can expect that once an AI of our creation becomes qualitatively more generally intelligent than us, it is likely to improve its own mind and get smarter and smarter, in ways that the human mind can't comprehend.

The limitations of our own minds are fairly obvious ... to name just a handful:

- our short and long term memories are both badly limited
- we need to use external tools like books and computers and calculators and wind tunnels and so forth to carry out cognitive operations that, for future nonhuman minds, are likely to be immediate and unconscious

- our ability to communicate with each other is horribly limited, reduced to crude mechanisms like arranging sequences of characters to form documents (as opposed to, say, telepathic communication, which would be easily simulable between minds living on digital computer systems).
- we're miserable at controlling our own attention, so that we regularly fail to do the things we "want" to do, due to lack of self-control (i.e. we have poorly aligned goal systems).

Rather often, when thinking about a math or science problem, a scientist comes up with an answer after years of thought and work - and the answer seems obvious in hindsight, as though it should have been clear right from the start.

The reason the answer wasn't clear right from the start is that humans, even the cleverest ones, aren't really very intelligent in the grand scope of things.

For transhuman AI minds, these intellectual problems that stump smart humans for years will be soluble instantaneously — and the things that stump them will be things we can't now comprehend.

# Why Bother to Build Such Things?

One may wonder why to create these minds? If they will evolve into something we can't understand, why bother — what use are they to us? Why not just create narrow-AI servants? Even if the narrow AI systems can't help us quite as effectively as superhuman AIs, they can probably do a fairly good job.

But life isn't just about one's own self. Just as there's intrinsic value in helping other humans, there's intrinsic value in helping these other minds — to come into existence.

These transhuman minds will experience growth and joy beyond what humans are capable of. Most likely, once the possibility exists, the vast majority of humans will choose to (rapidly or gradually) transform themselves into transhuman minds, so as to achieve a greater depth and breadth of joy, growth and experience. But the choices of those humans who want to remain human should also be respected.

## What About the Risks?

There are risks in creating superhuman minds — risks to humans, and also risks to these minds themselves (although the latter are harder for us to understand at the moment).

But, Cosmism is not about faint-heartedly fearing growth because it comes with risks. Growth always comes with risk, because it involves the unknown.

Cosmism is about managing the risks of growth intelligently, not avoiding them out of fearfulness and conservatism.

Transhuman AGI? Bring it on!

Design it with proper values in mind, then bring it on — and may the joy, growth and freedom continue!

## Brain-Computer Interfacing

Another powerfully transformative technology on the way is brain-computer interfacing — BCI — the ability to plug computer hardware into our brains.

It's hard to predict the precise transformations this will enable, but they are sure to be dramatic ones.

Adding new sense organs and actuators is just the start. Seeing at night ... gathering data directly from weather satellites or a car's or plane's sensor system ... driving or flying as if one were operating one's own body....

Adding new cognitive abilities — like instantaneous cognitive access to Google, calculators, Mathematica, Wikipedia, etc. — will transform the way we think.

Brain-to-brain linkages will allow a form of "telepathic" communication without any psi power required, and transmitting both emotions and thoughts. Mutual understanding in a whole new way.

And what about the potential for enhanced self-control?

Want to focus on studying for that test? Just program your brain-computer interface device to focus your attention properly for the next 8 hours.

Tired of being attracted to the wrong romantic partners? Program your interface to stimulate your love centers when around folks you *want* to be attracted to. Etc.

The most exciting applications, in all probability, will be the ones we haven't thought of yet, and won't imagine till they're here.

And then there's the possibility of using brain-computer interfacing to interlink human brains into a global distributed human/digital computing system ... a global brain mindplex ...

## Global Brains and Mindplexes

It's not necessary to think about humans and AIs with an "Us versus Them" mentality ...

One very real possibility is that humans, narrow AIs and AGIs in some sense merge together into a collective intelligence — an emergent "global brain" (which might of course extend beyond the globe as such, assuming humans or AIs voyage into space).

Humans could bear a number of different relationships to this global brain.

Via jacking into the global brain with brain-computer interfacing hardware, people could quite directly partake of it, perhaps sacrificing much of the individuality we currently associate with being human, but gaining a feeling of oneness with a greater mind, and the capability to share with other humans with a richness not possible in conventional human mind-states, and to partake of processes of thought and feeling beyond human ken.

Or, for those who favor retention of greater individuality, connecting to the global brain could be more like using the Internet today — but an order of magnitude more pervasive. What if your computer, your cellphone, your car, your service robot, all the appliances in your house — were all part of the same global nonhuman intelligence? You wouldn't need a cranial jack to be locked into the global brain and have your psyche and self adapt to it. You'd make your free choices and have your autonomy — but everything you did would be influenced and conditioned by the global brain.

Assuming a free society, interacting with the global brain would be optional — but nearly everyone would take the option, just as so many other highly convenient, inexpensive technologies have been adopted by nearly all people given the chance.

# Is the Global Brain Already Here?

Society now is already a "global brain" in a sense —
but it seems to lack the emergent reflective, deliberative
consciousness that humans have.

What if we create AGI systems that scan patterns in
the Internet as a whole, and attempt to guide global
thought trends by inserting information appropriately on the
Net for people to act upon? In this manner, we could
supply society with a theater of reflective awareness —
thus making a Global Brain with more of a purposeful,
explicitly goal-oriented coherence, so it would be
something more like a human mind. This was the core
goal underlying the company *Webmind* (originally named
*Intelligenesis*) that I founded in the late 1990s: the idea
was a bit ahead of its time then, but the world is gradually
catching up....

This may be the context in which superhuman AI
develops ... and in which some humans decide to
transcend legacy human awareness and become
something smarter and broader.

## A Global Brain Mindplex?

The achievement of a true global brain (with an emergent, global theater of reflective consciousness) without the abolition of human individuality would constitute an example of a mindplex — a mind that has reflective consciousness on more than one level ... a reflectively conscious system some of whose component parts are also reflectively conscious systems.

We don't know much about mindplexes — they don't exist yet on this planet, so far as we know; and their dynamics will doubtless involve many strikingly novel phenomena.

# Morphogenetic Fields and the Collective Unconscious

Carl Jung famously wrote of the Collective Unconscious: a transpersonal pool of archetypal patterns binding us all, giving us abstract, emotionally, spiritually and interpersonally meaningful shapes which we then flesh out into individual concepts.

He saw this as tied into psi phenomena, and as reflecting an order of reality beyond the physical. His evidence was anecdotal, consisting partly of the observation of many so-called "synchronicities" — coincidences in everyday life that seemed extraordinarily improbable according to the traditional physical understanding of the world.

Rupert Sheldrake put forth the related idea of a "morphogenetic field" — a kind of life-field containing structural and dynamical patterns that provide living beings with archetypes to flesh out as they grow and adapt. Morphogenetic fields have been proposed to play a role in epigenesis (providing infant plants and animals with shape-templates to follow as they develop) and also in learning (providing a way for one population of organisms to transmit learned knowledge to other distant organisms, without aid from any traditional physical causal mechanism).

The morphogenetic field bears striking resemblance to the Qi energy field posited in traditional Chinese spirituality and medicine.

There is great resonance between the morphogenetic field concept, and the notion of "pattern space," of the network of shifting, overlapping, inter-emerging patterns as a fundamental domain of being.

In the morphogenetic field as posited, two physical entities with similar patterns, may be conceived as having a sort of "channel" between them, along which further patterns may flow. This notion provides a dramatic extension of the "tendency to take habits" principle from the abstract domain of Thirdness into the physical world specifically.

The Jungian collective unconscious could be interpreted as the implication of the morphogenetic field for human brains.

## Physical, Not Just Metaphysical, Hypotheses

The morphogenetic field and collective unconscious might be interpreted in two broad ways:

- as descriptions of underlying pattern-space, of the transpersonal world out of which individual minds and the physical world both crystallize

- as descriptions of underlying pattern-space, but also of scientifically understandable phenomena that occur in physical reality (reflecting these underlying pattern-space phenomena)

Their validity in the first sense is plain. Their validity in the second sense appears to me a currently open question.

But it is the second sense that Jung and Sheldrake proposed: this is the bite and the drama of their ideas.

As with psi phenomena, Cosmism embraces the collective unconscious and the morphogenetic field as real scientific possibilities. However, whereas there is dramatic scientific evidence in favor of psi phenomena, the same cannot currently be said for the collective unconscious or morphogenetic fields. It is abundantly clear that collective unconscious and morphogenetic fields have a reality from the subjective, interpersonal and transpersonal perspectives, but in my view it is currently ambiguous whether they have a reality from the "physicalist", third-person perspective.

To the extent that the morphogenetic Qi energy field does have physical reality, it may be (at least partially) understood as a means by which pattern flows directly from one region of physical space to another, without making use of the physical dynamics used for known mechanistic forms of signaling. That is: it may be conceived as a *direct flow-through from pattern-space into physical space*.

There is particular conceptual resonance between morphogenetic fields and certain forms of psi. For instance, consider the phenomenon via which identical twins sometimes have an intuitive, "telepathic" knowledge of major events in each others' lives. Morphogenetic fields provide a natural explanation: the similar patterns in the twins' minds create a "channel" through which other patterns flow.

# Morphogenesis of the Technological Singularity

If morphogenetic fields do have a physical reality, this has striking implications for the possibility and nature of a technological Singularity. It suggests that our various individual and group attitudes on the Singularity could play a major role in its realization. If the human species confronts the Singularity with archetypes of doom, destruction and conflict, this may impact what actually occurs, in a negative direction. If the human species confronts the Singularity with archetypes of wonder, joy and cosmic harmony, then this may also impact what actually occurs, in a positive direction.

Of course, attitudes toward the Singularity will likely influence its nature even without any strange mind-fields intervening, via well-known psychosocial mechanisms. Physically impactful morphogenetic fields would "merely" make the influence more dramatic!

## The Strengths and Limits of Science

The transhuman technologies whose implications occupy so many of these pages, are all coming about as a result of the human institution of *science*.

Science is a wonderful thing ... I was taught in my early youth by my grandfather, a physical chemist, to revere it above other human institutions — as a way of humanity finding a kind of truth and purity not present in most other human pursuits.

But still — we mustn't exaggerate its scope and its power.

Science, as we currently conceive it, is based on finite sets of finite-precision observations. That is, all of scientific knowledge is based on some finite set of bits, comprising the empirical observations accepted by the scientific community. All the empirical knowledge currently accepted by the scientific community as the basis of scientific theory, could be packed into one large but finite computer file.

To extrapolate beyond this file, this bit-set, some kind of assumption is needed. Or, to put it another way: some kind of "faith" is needed.

David Hume was the first one to make this point really clearly, a couple hundred years ago ... and we now understand the "Humean problem of induction" well enough to know it's not the kind of thing that can be "solved." As Hume noted, just because we have observed the sun rise 5, 50 or 500000 mornings in a row, doesn't justify us in assuming it will rise the next morning. This prediction, this "induction," rests not only on our prior observations, but on some kind of assumptive theory.

The Occam's Razor principle tries to solve this problem — it says that you extrapolate from the bit-set of known data by making the simplest possible hypothesis. I.e., it says that *patterns* (defined as representations-as-something-simpler) *tend to continue*. This leads to some nice mathematics involving algorithmic information theory and so forth. But of course, one still has to have "faith" in some measure of simplicity!

So: doing or using science requires, in essence, continual acts of faith (though these may be unconscious and routinized rather than conscious and explicit).

This doesn't make science a bad thing, and it doesn't detract from science's incredible power and usefulness.

The body of human scientific knowledge is best viewed as a kind of living organism, as a mind unto itself — a mind which grows, makes choices, and experiences joy as it confronts and creates ongoing surprises. Being a scientist is largely about communing with this mind — about fusing one's individual mind with the greater collective mind of science. This fusion, though never quite complete, can have profoundly transformative effects upon the individual self, sometimes resulting in individuals who have very little self-sense and are primarily operative as subsets of the greater mind of science.

The mind of science is far more powerful than any individual human mind. Yet it is not absolute; it is not always "right"; and it does not escape the need to base its judgments on some raw assumptions, some assumptions that do not emerge directly from empirical observation or mathematical derivation.

The extent to which the mind of science will survive the transcension of humanity seems quite unknown. Perhaps the separation of science from other modes of life, growth and inquiry is an epiphenomenon of the human cognitive architecture. Perhaps science will merge together with other sorts of pursuit, in the psychology and community of superhuman minds. In this sense, the mind of science, like the individual human mind, could contain the seeds of its own transcension.

One thing that points to a possible "trans-science" is the study of consciousness. It seems that to really understand conscious experience, will require some new sort of discipline — something bringing together subjective experience, shared social experience, and scientific data-gathering and relation-forming. Francisco Varela and David Bohm were two scientists who explicitly worked toward the formation of such a discipline — and dialogued with the Dalai Lama and other spiritual seekers about the idea.

As Cosmists we must respect the pattern of science, which has brought us so far. But we must also be open to its transcendence, in ways we're now unable to foresee.

## Language and Its Children

Language is a wonderful thing — without it I wouldn't be able to have these thoughts, let alone communicate them to you.

It's so wonderful that we sometimes forget how measly it is: setting aside intonation and gesture, which only exist in spoken language, it purely consists of the arrangement of a finite set of characters in various finite lists.

How amazing that these finite arrangements, these encodings, can serve as such a powerful tool for communication and coordination among intelligences!

But language also has its limitations, and our future may hold some better tools.

## Lessons from Lojban

One of the most interesting languages on Earth is Lojban, which is a written and spoken language for everyday informal communication that has a syntax and semantics based on predicate logic (a form of mathematical logic). Lojban is precisely parse-able in the same way as a computer programming language, yet can be used to communicate everyday things between people.

*mi cu tavla do la lojban*

("I speak Lojban to you")

One of the lessons Lojban has to teach us is where language gets its communicative power. The difference between Lojban and mathematics is that, in Lojban, even though the syntax is mathematically defined and the general semantic relationships between elements of a sentence are mathematically defined, the relationships between words and the world are left informal.

Lojban attempts to make words as precise as possible — for instance, instead of a vague word like "write" there are separate words for "authoring" a book versus "scribing" a book (i.e. typing it or writing it out by hand). But there are limits. Ultimately, even in Lojban, the significance of a word in a context has to be figured out via nonlinguistic reasoning or intuition.

What this tells us is: language exists to channel and direct nonlinguistic understanding among minds with a shared understanding of a commonly perceived reality. It doesn't exactly "describe" reality — it serves as a tool that members of a community can use to coordinate and channel their shared, internal descriptions of reality.

## Limitations of Language

And language has profound limitations. Not all aspects of shared understanding can effectively be channeled through language.

Even accounting for the power of love poetry — still, love is best communicated nonverbally. (Language seems to do a better job of communicating love when coupled with music; hence the popularity of love songs.) So are many other emotions and aspects of interpersonal relationships.

"Mathematical maturity" — the ability to approach complex math proofs in an appropriate way — is best communicated via example and via cooperation in theorem-proving, rather than by linguistic explanation.

# Beyond Language

It seems likely that as transhumanity unfolds, language as we now understand it will become a thing of the past. Direct mind-to-mind transmission of information will be the most likely replacement.

Different minds have different internal vocabularies, and so there may emerge "intermediary minds" serving as common conceptual vocabularies, so that two very different minds can communicate by exchanging thoughts via an intermediary. One could think of this as a kind of "Psynese" language, but it would be very different than anything we call "language" today.

Writing a book like this makes me acutely aware of the limitations of language. How much more fun, and useful, it would be if I could transmit these thoughts more directly into your mind!

# The Strengths and Limits of Mathematics

Mathematics is one of our most powerful and perplexing inventions.

From one perspective, it's just a system for making various sequences of marks on pieces of paper (or computer keyboards, etc.). A mathematical system tells you which sequences of marks are "allowed" or not; and then the doing of mathematics consists of figuring out which sequences of marks are allowed in a certain system. Sort of like a language where the basic grammar principles are known, but are so tricky that it's a hard puzzle to figure out which sentences are grammatical or not.

The purpose of this "mathematical marks game" is that some people find it beautiful and entertaining ... and that people know how to correlate some of the marks with actions and perceptions in the world, thus allowing mathematics to be used in physics, biology, sociology and so forth.

From another perspective, mathematics describes realities beyond the one we live in. For instance, there are various theories of huge infinite sets. The "existence" or otherwise of these sets can never be validated by science, because science ultimately has to do with finite sets of finite-precision data. But, mathematical theories involving these sets may nevertheless be very useful to science.

And the communication of information about these infinite sets via language is an interesting thing — because language, like science, has to do with finite sets of data (finite texts composed of characters drawn from a finite alphabet).

But if we remember that language doesn't encapsulate knowledge — it rather serves to channel shared understanding — then this isn't so mysterious. If we human minds have shared understanding of these infinite sets, then language can serve to coordinate and channel this shared understanding. This is what it feels like is happening when mathematicians discuss abstract mathematics.

## Can Digital Computer Programs Understand Mathematics?

Humans' apparent ability to intuit infinite sets makes things interesting for the AI theorist — because, what would it mean to say that a mind implemented as a finite, digital computer program could enter into a shared understanding of an infinite set?

Some AI theorists (for instance, Selmer Bringsjord) argue that digital AGI programs are only capable of understanding infinite sets indirectly, as certain finite arrangements of symbols — whereas we humans can apprehend them directly

This is possible, but I'm skeptical.

Rather, I think what's happening here is well-understood in Peircean terms as a confusion between Firsts, Seconds and Thirds.

Infinite sets have their own unique Firstness ... but in their Thirdness (not their Firstness or Secondness) they are reducible to symbol manipulations, to sequences of characters.

I doubt that, when we humans intuit infinite sets, our brains are doing something fundamentally different from when we intuit the number "5", which mathematics models as a finite set.

It seems quite feasible that advanced digital computer programs will, like humans, be able to experience a Secondness, in which their own Firstness collides with the Firstness of infinite sets.

One wonders if some future discipline might weave aspects of current mathematics into other aspects of experience. If a future science of consciousness brings subjectivity and objectivity together in some novel way — will it come along with some allied novel discipline binding the formalism and experience of infinity?

# Reason and Intuition

"Reason" is one of the most powerful tools human culture has developed. It has various forms, including sophisticated verbal argumentation (like law, philosophy, and the analytical portions of "humanities" generally), and the numerous species of mathematical logic.

In plotting our course toward the future, it's important that we try to be as reasonable as we can. Thinking carefully is a wonderful way (though not the only way: e.g. meditative disciplines help with this too) of avoiding being pushed around by the more animalistic portions of our brains. And reason is the best tool we have for figuring out likely conclusions from our observations and assumptions.

But overvaluing reason would be just as foolish as ignoring it. Reason is a powerful tool but in some contexts it is so inefficient it is impractical to apply. It is especially inefficient at arriving at conclusions based on massive amounts of heterogeneous data. Sometimes the most reasonable thing to do is to set aside detailed reasoning about a certain matter and make a judgment by intuition!

And reason can never be a complete solution to understanding the world, because reasoning always relies upon certain assumptions — which must come from somewhere besides reason.

## Intuition

As human beings, we need to rely on reason plus intuition — the latter being a crudely-defined shorthand for "certain human brain/mind processes that synthesize some processes in the theater of reflective awareness and some outside it, aimed at arriving at solutions to problems based on holistically integrating all the information available to the mind, or at least a large percentage thereof."

Intuition's conclusions may not always be easy for us to justify by reason, in practice. A question is whether, in principle, given enough space and time resources and enough visibility into the unconscious mind, one could always justify a good intuitive conclusion based on rules of sound reasoning. I suspect this is true — but even if so, it's not a very helpful thing, because in practice we don't have arbitrarily much space and time resources to carry out this sort of experiment, and we also don't have the capability to suck all the contents of a human's unconscious into some reasoning engine's theater of reflective awareness.

As a scientist, I have great interest in understanding the workings of reason — for instance, I do research on the application of probability theory and formal logic to reasoning about fuzzy everyday events. But at the same time, when I do science, I rely on a mixture of reason and intuition just like everybody else!

## Will Intuition Go Obsolete?

Perhaps future AI minds will have greater capability to reason than us — associated, perhaps, with much larger and more flexible theaters of reflective awareness. But even so, I strongly suspect they will still need a mixture of reason with some form of intuition. I suspect that reason will always be a resource-intensive, complex approach to solving problems that depend on large, heterogeneous pools of information — so that it will always be supplemented with other methods with different strengths.

But I freely admit that this is conjecture — perhaps new forms of reason will be admitted that don't have the shortcomings of human reason. Perhaps future minds will solve everything using some transhuman form of logic, even the choice of where to place their little toe when they walk, and the choice of which key to play next on the piano (or whatever analogous choices they have to make).

## Science-Friendly Philosophy

Doing science, mathematics and engineering — just like other aspects of living life — relies on a constant stream of acts of faith, which can't be justified according to science, mathematics or engineering....

If one views science as operating according to Occam's Razor — the choice of the simpler hypothesis — then these "acts of faith" have a lot to do with basic assumptions about what feels simpler.

But how are these "acts of faith" organized? How do they interact with each other?

There are various systems for mentally organizing one's acts of faith.

Religions are among these systems. But religions are quite detached from the process of doing science, math or engineering. Adopting religion as a primary method of organizing one's acts of faith makes thinking about science on a profound level awkward.

*(For the rest of this chapter, I'll use "science" as an abbreviation for "science, math and engineering", just to avoid overly long and tedious sentences.)*

It seems sensible to think about philosophical systems — i.e. systems for organizing inner acts of faith — that are *intrinsically synergetic with the scientific process.* That is, systems for organizing acts of faith, that

- when you follow them, help you to do science better
- are made richer and deeper by the practice of science

Now, one cannot prove scientifically that a "science-friendly philosophy" is better than any other philosophy. Philosophies can't be validated or refuted scientifically.

So, the reason to choose a science-friendly philosophy has to be some kind of inner intuition; some kind of taste for elegance, harmony and simplicity; or whatever.

One prediction I have for the next century is that science-friendly philosophies will emerge into the popular consciousness and become richer and deeper and better articulated than they are now.

Because, even as science becomes ascendant over traditional belief systems like religions, people still need more than science ... they need collective processes focused on the important philosophical questions that go beyond the scope of science.

So, my prediction is that we are going to trend more toward philosophical systems that are synergetic with science, rather than ones that co-exist awkwardly with science.

Cosmism is one example of a philosophical system of this nature!

There's nothing extremely new about the concept of science-friendly philosophy, of course.

Plenty of non-religious scientists and science-friendly non-scientists have created personal philosophies that don't involve deities nor other theological notions, yet do involve meaningful approaches to personally exploring the "big questions" that religions address.

Among the many philosophers to take on the task of creating comprehensive science-friendly philosophical systems, perhaps my favorite is Charles Peirce. But Peirce was writing at the turn of the 20th century ... he lacked the insight into science, math and technology that we now have.

Cosmism is intended as a science-friendly philosophy that is adequate to carry us to, and maybe through and beyond, a Singularity or another sort of transcension event.

## Art Is the Ultimate Occupation

As the advent of advanced technology makes labor unnecessary for advanced intelligences, one aspect of human life that will gain increased rather than decreased importance is *art*.

By "art" I don't mean specifically the creation of paintings, drawing, plays, symphonies, dances, novels and so forth — these are great, but they're just particular examples.

I mean the *creation and sharing of new patterns purely for the sake of having these patterns appreciated by one's own mind and others' minds*.

Once the need for humans and other advanced intelligences to labor for sustenance is eliminated, what will be left for minds to do is primarily to **create and appreciate art**.

"Art" may involve building baby universes or new AI systems or 9-dimensional meta-multiversal movies ... or scientific data-sets or theories or mathematical theorems ... or perfecting one's own array of mind-states ... or drawing pictures of trees on paper ... all of the above and many more! (Let infinity flowers bloom...!)

Where art is concerned, the main point is not the particular medium or even the product but rather the motivation.

For example: Now people do science in part because they get paid for it, in part out of a desire to help in the process of creating new technologies to make peoples' lives better, and in part out of a desire to understand and create beautiful knowledge. Once scarcity and suffering are largely palliated, only the last motive — which is essentially artistic — will remain.

*The artists shall inherit the Cosmos.*

## Creative Nihilism

One meme that has gotten an unjustly (and calculatedly) bad rap is "nihilism."

Typically taken to signify "believing nothing has any meaning or value," it originally meant something quite different.

Dostoevsky, a genius writer and a profoundly religious man, parodied nihilism mercilessly and hilariously in his novels, interpreting it in the above manner.

But many of the Russian nihilists of the mid-1800s (for instance the great mathematician Sofya Kovalevskaya, perhaps my favorite female mad scientist) took it to mean, rather, "believing that nothing has any *absolute* meaning or value" — a radically different thing!

Cosmism advocates nihilism in the latter sense — which I call *creative nihilism.*

*Take nothing for granted!*

As the bumpersticker says: *Question authority!*

Or in the words of William Burroughs: *Nothing is true! Everything is permitted!*

But Burroughs didn't mean "nothing is true at all." He meant "nothing is absolutely true."

And he didn't mean "every activity and idea is permitted in every context." Some valuable contexts, clearly, are defined by what they rule out.

He meant "nothing is a priori ruled out. There is total freedom to explore."

Cosmism requires creative nihilism: anything else stifles growth and choice, restricting avenues for joy.

*Long live creative nihilism!*

## The Ethics of Creating Transformative Technologies

Is it right to create radical new technologies when they are potentially dangerous?

Shouldn't we prioritize the survival of our species, rather than taking risky gambles on new technologies that could lead to great things but could also lead to destruction?

Shaping the future involves a host of difficult balancing acts, indeed.

## The Proactionary Principle: Weigh the Costs of Action versus the Costs of Inaction

If the human world were a well-organized, peaceful place, in which some benevolent Central Committee of Technology made centralized decisions about what technologies to explore at what paces — then, almost surely, it would make sense to manage our development of powerful technologies very differently than we do today.

But that's not the world we live in. In our present world, multiple parties are working on advanced, potentially radically transformative technologies in diverse, uncoordinated ways.

Many of these parties are working with an explicitly military goal, oriented toward creating advanced technology that can be used to allow one group of humans to physically dominate another.

In this context, there is a strong (though not unassailable; these are difficult issues!) argument that the most ethical course is to move rapidly toward beneficial development of advanced technologies ... to avoid the destructive (and potentially species-annihilating) consequences of the rapid development of advanced technologies toward less beneficent ends.

An extreme form of this position would be as follows:

*We humans are simply too ethically unreliable to be trusted with the technologies we are developing ... we need to create benevolent artificial general intelligences to manage the technology development and deployment process for us ... and soon, before the more monkey-like aspects of our brains lead us to our own destruction.*

Or in other words:

### Do we need an AI babysitter?

# Existential Risks

There is a group (I'm on their Board, but so far not heavily involved) called the Lifeboat Foundation that exists to look out for "existential risks" — things that threaten the survival of the species. This is a worthy pursuit — but at the moment, it's very difficult for us to rationally assess the degree of risk posed by various technologies that don't yet exist, or exist only in immature form.

One macabre theory for the apparent lack of intelligent life elsewhere in the Cosmos is the following: on various planets in the galaxy, as soon as a civilization has reached the point of developing advanced technology, it has annihilated itself.

A less scary variant is that: once a civilization reaches advanced technology, it either annihilates itself or Transcends to some advanced mind-realm where it's no longer interested in sending out radio waves or gravity waves or whatever, just to reach civilizations that are in the brief interval of having reasonably advanced tech but not yet having reached Singularity.

## Selective Relinquishment

Ray Kurzweil, among others, advocates "selective relinquishment," wherein development of certain technologies is slowed while advanced technology as a whole is allowed to accelerate toward Singularity. This seems what is most likely to happen. The outcome cannot be predicted with anything near certainty.

It seems relevant to quote the famous Chinese curse: "May you live in interesting times."

Which from a Cosmist view is — of course — closer to a blessing than a curse! But traditional Chinese values favor stability — whereas Cosmist favors growth, while also respecting the importance of preserving the better aspects of the past.

Certainly, we must approach the unfolding situation with ongoingly open hearts and minds — and appropriate humility, as we are each but a tiny part of a long evolutionary dynamic, that extends far beyond our current selves in both past and future.

But there is also cause for activism. The future is what we will make it. Sociotechnological systems have chaotic aspects, so small individual actions can sometimes make dramatic differences. There may be opportunities for any one of us to dramatically affect the future of all of us.

## Does Cosmism Advocate Human Extinction?

Cosmism, as I conceive it, is about seeking a positive life based on actively seeking increasing knowledge about the Cosmos in all its aspects.

Joy, growth and choice and all that!

But what does this mean about us (smelly, hairy, violent, sex-obsessed, chaotically creative and cultured, beautiful, loving and malevolent,...) *people*, in particular?

Of course, you could have all these glorious, abstract-sounding values preserved without any humans around.

But the existence of humans — in spite of all our imperfections — certainly doesn't contradict joy, growth and choice. Indeed, the forcible abolition of humans would be a rather strong violation of the value of choice.

What Cosmism *encourages* is not the abolition of humans, but the transformation of humans into something more joyful and more splendidly growing than current humans — guided not by force but by human intentionality.

Cosmism *does not* encourage the forcing of transformation or transcendence or transhumanity on humans whose choice is otherwise.

Cosmism *does* advocate not allowing those who choose to remain "legacy humans" to diminish the joy, growth and choice of others — most likely there will always be some balancing to be done, as maximizing all three of the "joy, growth and choice" values may not be possible given the constraints posed by the universe.

## Hypothetical Tough Choices

Hypothetically one can construct scenarios where there is a clear, crisp choice between, say,

- A static, depressing, fascist world dominated by humans
- A joyful, growing, freedom-ful world without humans

and then ask which one is preferable.

The Cosmist answer is obviously: the latter.

In Cosmism, humans are valued as sentient beings and complex pattern-systems — but they're not viewed as uniquely important, and if it happened that the persistence of humanity violently contradicted higher, broader values, then the values would win.

But this kind of scenario seems extremely unlikely to occur — for one thing because humans are just not going to be that powerful compared to transhuman minds we will create (or that our creations will create, etc.). It seems unlikely humans will have the power to significantly perturb the joy, growth and freedom in the future universe, even if they wanted to. My gut feeling is that once we have transcended the legacy human condition, these artificial dichotomous situations are going to look very silly in hindsight.

Someone asked me, recently, the following question:

*Hypothetically, if there were a situation in which you knew that the development of AI would directly harm a massive amount of people would you decide to end your work or keep going?*

I won't repeat my whole answer here but the core of it was as follows:

*If a path to AGI is leading in the direction of "necessary" massive destruction, it's probably a suboptimal path, and a better path to AGI can be found.*

**Obsolete the dilemma!**

# Shouldn't We Seek to Guarantee the Ongoing Welfare of the Human Race?

At the moment, my gut feeling (which could change as we all learn more about these issues) is that any kind of *guarantee* of human well-being unto eternity post-Singularity, is going be bloody hard to come by.

It seems more feasible to me that one could come close to guaranteeing a peaceful "controlled ascent" for those humans who want to increase their mental scope and power gradually, so that they can experience themselves transcend the human domain.

A more important, statement, perhaps, is that early-stage AGI scientists are likely to help us understand these issues a lot better.

But it's important to recognize that fundamental growth inevitably involves risk. Growth involves entering into the wonderful, frightening, promising unknown. In this kind of situation, guarantees are not part of the arrangement....

# The Quest for Unifying Laws of the Cosmos

A popular meme in modern culture is the search for Universal Laws of the Cosmos ... for some fundamental, unifying equation.

The possibility can't be denied — but even if such a thing does exist, it seems extremely likely that we're extremely far away from being smart enough to find or understand it.

As our intelligence increases, and as we create new intelligences with power and scope far beyond that of legacy humans, it seems likely that new aspects of the universe will get discovered, from the perspective of which our current understanding will seem nearly as silly and limited as the ancestor-worship-based world-views of pre-civilized tribes.

By all means, let us seek unified field theories, unified equations and other unified understandings.

We have gained many insights from this quest so far. General relativity has taught us that space and matter are intimately interdependent, and best thought of as creating each other. Quantum physics has taught us that even from the perspective of scientific experimentation, the domain of physical reality must be thought of as, in a sense, observer-dependent. Neuroscience and genetics have revolutionized our self-understanding; a few aspects of this surfaced above in the discussion of free will, consciousness and so forth.

But let's not fool ourselves that these understandings are terribly likely to retain their appearance of grand scope after we enhance our intelligences and understand more and more of the world!

## The Complex Cosmos

Another version of the quest for a unified scientific understanding of the Cosmos is the *science of complex systems* – which seeks broad principles describing the structure and dynamics of complex self-organizing systems in various domains, including physics, chemistry, biology, sociology, economics ... and maybe even the Cosmos as a whole.

Modern complexity science is even more primitive than modern physics, but yet, it does have some suggestive lessons for Cosmism.

If we view the universe as a network of patterns among patterns among patterns ... and we adopt the goals of joy, growth and choice — then we arrive at the question: *What sorts of pattern networks will generally lead to greater joy, growth and choice*?

It would be foolish to seek any general answer to this question just now, even if our complexity science were more advanced. We humans understand so little compared to the scope of the Cosmos. But a few potentially relevant principles have emerged from complexity science so far.

**Connectivity!** It's important for all the parts of a system to be interconnected; this allows adaptiveness, robustness, and the emergence of complex patterns

**Differential connectivity**: many productive and robust complex systems seem to have a few elements that are much more richly connected than the rest. Mathematical phenomena like "small world networks" explore this aspect in detail.

**Neither too orderly nor too chaotic** — many complex systems seem to have a self-regulatory mechanism that keeps them somehow on the boundary between order and chaos; so that they avoid falling into ruts, but also avoid too frantically disrupting the complex patterns they've discovered and the ones they're in the midst of forming

**Hierarchical structure**: with simple patterns combining to form more complex ones, combining to form more complex ones, etc. Not all complex patterns are hierarchical, but complex hierarchical patterns are particularly easy to form and find, and seem to play a critical role in both the physical, social and mental worlds.

**Dual network structure**: many real-world hierarchies are associative structures as well, in the sense that entities nearby in the hierarchy also have a lot of other similarities not obvious from the hierarchy itself.

For instance if we view the physical world as a hierarchical space-time structure, we find that spatiotemporally nearby entities also seem to be related in other ways.

And if we look at the human mind as a hierarchy, we see that entities nearby in the "natural concept hierarchy" (say, cat and dog, or table and chair) tend to be associated in a rich variety of ways.

Ultimately this is a consequence of Peirce's "tendency to take habits" — the habitual pattern of organization embodied in the hierarchy, tends to correlate with a host of other habitual association patterns.

**Reflexive structure and dynamics**: many complex patterned systems are good at recognizing patterns in themselves.

By recognizing a pattern in itself, and then embodying this pattern as part of itself, the system grows more self-aware, richer and more intelligent. This is how self and reflective consciousness work — and it's also intrinsic to *life* more generically.

There is a traditional in theoretical biology, founded by Humberto Maturana and Francisco Varela, which focuses on the critical role reflexivity plays in biological self-organization. They use the term *autopoiesis* – self-poetry; self-construction – and trace out the complexity and beauty of its operation at multiple levels, including the cell, the organ, the organism and the ecosystem. A biological entity, as part of its very aliveness, recognizes patterns in its own structure and dynamics and embodies them in its actions and self-modifications.

Qualitatively, all these seem to be general principles spanning many complex systems in psychology, biology, chemistry, physics, sociology, economics, and so forth — and so it is tempting to conjecture they have some general meaning, going beyond the scope of the world we humans now know. (And the above is far from a complete list, but just a sampling of some complexity science principles that have leaped out at me as particularly significant in the context of my own work and thinking.)

Contemporary science and mathematics groks these ideas only in a quasi-rigorous way — today's complexity science is a melange of highly specific rigorous theories, bound together by some rough high-level concepts. And even a rigorous science/math understanding wouldn't necessarily carry over into the transhuman domain.

But still, as we reshape ourselves and our world, it pays to keep the principles of complexity in mind.

The Internet seems to be evolving into something fulfilling these principles of complex systems, which is an interesting and positive sign, and something to keep our eye on as the Global Brain emerges.

Similarly as we architect and interconnect AGI systems, the principles of complexity may usefully guide us.

And who knows, we may even see these principles in post-Singularity spaces we can't currently envision — it wouldn't be shocking if basic principles of self-organization transcended our little human corner of the Cosmos.

## Separateness, Togetherness and Evil

One of the biggest dilemmas of classical philosophy is the so-called "problem of evil."

That is: why does bad stuff exist? Why is there pain? Why is there torture?

This was rather mystifying to those medievals who believed in an all-knowing, all-powerful, all-merciful God. After all: if God is so great and sees everything, and if he cares about us humans, why does he let us get eaten by sharks, or let us peel each others' skin off, etc.? What about starvation, natural disasters?

The classical answer is that it has something to do with free will: God (or the Devil, or someone) gave us free will, and with that came suffering, because God couldn't eliminate our suffering without eliminating our freedom also. Because in some way, our suffering is self-caused, caused by our own chosen actions.

If you've read the previous parts of this Manifesto, you know I don't place much stock in free will nor devils ... but even so, I do think there is something to the "trade-off" aspect of this classical answer.

But I tend to think about it more in terms of the trade-off between *separateness* and *togetherness*.

Going way out on a metaphysical limb, my suggestion is that: If one wants to have a universe with a bunch of separate entities, rather than just one blurred-together lump of indivisible being ... then one is going to have bad stuff, one is going to have pain and woe and all that in some form or another.

*Have you ever said Yes to a single joy? O my friends, then you said Yes too to all woe. All things are entangled, ensnared, enamored* — Friedrich Nietzsche, in *Thus Spake Zarathustra*

All pain, I suggest, is ultimately rooted in pain of separation. The emotional experience of pain arises from signals informing an organism of potential dangers to its ongoing existence as a separate, autonomous entity. All pain and "evil" is ultimately a result of the existence of separately bounded entities.

But — stepping even further out on the precarious metaphysical limb — this raises the question of *why separately bounded entities should exist at all?* Why isn't there just one big happy, fuzzy, cosmic moment?

(Yes, way back in the introduction to this text, I promised to keep things practical and not delve too extravagantly into metaphysics. But this is a brief digression which I feel quite important, so I hope you'll forgive me!)

My intuition on this is a simple one....

Earlier, following the old-time psychologist Paulhan, I suggested that a core aspect of Joy is "unity gain" — the feeling of *separate things coming together* ... the increase of unity and patternment.

And unity gain between minds with selves is nothing more or less than — our old friend *love* ...

Which brings us to an interesting conclusion: separateness, the cause of pain, is necessary so that joy, the feeling of increasing togetherness, can exist.

No separateness, no feeling of increasing togetherness.

*Separation exists to enable love.*

The crux of joy and love, then, is: *obsoleting the dilemma* of separation.

Too much abstract metaphysics, perhaps ... yet I can often feel the raw truth of this perspective in events in my everyday life. Maybe you can as well.

I'll close with a comment Chase Binnie made on the above text:

*I'm with you on this Ben. Much of my own suffering is caused by alienation or feeling disconnected. Connection to other people is the reconnection of us to ourselves.*

## Universe, Multiverse, Yverse

One of the gifts quantum physics has given us is a model of the universe very different from the standard one. I'm talking about the notion of a "multiverse" — a meta-space of multiple universes, wherein every time something happens, a "split" occurs and there is one branch of the multiverse where the thing happens, and one branch where it doesn't happen. Each branch of the multiverse is conceived to contain a multitude of universes ... and each of us is conceived to have "copies" in many, many multiverses.

So, for instance, there are branches of the multiverse in which I never allocated time to writing this Manifesto, and spent the time mixing down some of my music recordings instead. There are branches where I did write this Manifesto, but made it slightly less silly by omitting this sentence. Etc.

Borges envisioned a multiverse-like Cosmos in his story "The Garden of Forking Paths," but quantum theory made the notion more concrete, via positing it as a solution to the "quantum measurement problem." Roughly speaking, quantum theory comes out much simpler if one assumes we live in a multiverse rather than a conventional, single universe.

# Beyond the Multiverse

Quantum theory doesn't do it, but one can also go beyond the multiverse. One can imagine a family of multiverses, where each one is based on certain assumptions that span all their branches. For example, one could have multiple multiverses each obeying different laws of physics. Then one would have a multi-multiverse.

And why stop there?

Ultimately one arrives at a multi-multi-...-multi-verse, which I have given the name "Yverse,", defined as

$$Yverse = multi\text{-}Yverse$$

A Y-verse is, to put it crudely, a set of branches, each of which is a Yverse. This is different from an ordinary multiverse, each of whose branches is not a multiverse but an ordinary universe.

The mathematics and physics of Yverses remains to be elaborated!

## What Is this "Place" We Live?

Why am I bothering to throw these speculations at you?

Mainly to make the point that the Cosmos may be a much subtler and odder place than we currently understand.

Quantum theory, which currently baffles us so profoundly, may just be scratching the surface of deeper models and understandings of reality, which transhuman minds will comprehend.

We should not be so narrowminded and egomaniacal as to assume that our current understanding of the universe — or our current, wild-ass speculations — are anywhere near complete or correct.

## Building "Gods"

Arthur C. Clarke wrote, in the middle of the last century, *"Perhaps our role on this planet is not to worship God — but to create Him."*

Subtract the Earth-centric and Christian-centric phraseology, and the implicit confidence that there is such a thing as a preconfigured "role" for a species ... and you're left with a very Cosmist-friendly notion:

*A fascinating and potentially excellent strategy for moving along the Cosmist path is to create a superhumanly intelligent, powerful and benevolent entity — i.e., to "build a god."*

It could be a digital computer. It could be a quantum computer, or some sort of system yet unknown to us.

What would it do? It could solve our problems far more effectively than we can. It could invite us into its mindspace. Ultimately though, we just can't know what it would do, any more than a cockroach can predict the unfolding of human events like wars or elections.

Potentially the advance of joy, growth, choice, understanding and all that other good stuff could be massively accelerated and improved by having our own home-brewed god to help us.

Is there a risk here? Yes.

Do we thoroughly understand the risk-benefit tradeoffs involved in such a pursuit? Not at this point.

Will we ever fully understand these tradeoffs? Probably not, but we can surely grok them more fully than is the case right now.

A lot more study will be required before we'll know for sure if building a god is the best thing to do ... and if it is, at what stage in the development of our knowledge it's the best thing to do.

And, not coincidentally, a lot more study will be required before we'll know *how* to build a god in enough technical detail to do it.

Cosmism doesn't advocate jumping rashly into such an enterprise. It does advocate devoting significant resources and enthusiasm to the serious exploration of the possibility.

## We Versus Us

A few people, on reading an earlier draft of this Cosmist Manifesto, remarked that they felt it represented a kind of unrealistic, idealistic, hippy-ish idealism.

To paraphrase their complaint in my own words: "*It's all very well to daydream about universal love and everyone becoming happier, healthier and more intelligent and cosmic in a post-Singularity, post-scarcity, post-selfishness Cosmos — but the reality is, the world is full of people who want to oppress us, bomb us, imprison us, take our goodies and so forth. The real problem is defending our right to a better future against the bad guys and the bad social structures — not fantasizing about building gods and other improbably positive futures!*"

There is no denying the presence of oppressive, confrontational, dangerous forces in the human world today. And sometimes we must fight against them.

My friend and fellow AI researcher Hugo de Garis foresees a likely World War III between Cosmists and Terrans, where the latter are conceived as those who wish to avoid AIs outdoing humans, with a goal of making sure humans remain the dominant species on the planet.

Ray Kurzweil has retorted that any war between pro-advanced-technology and anti-advanced-technology forces is likely to be very short, due to the former possessing drastically superior weapons technology. There seems to be some truth to this beyond the witty-quip level, because current terrorist groups from low-technology cultures seem to have a very hard time mastering advanced weapons technology (which is a very good thing). And yet one cannot rule out the possibility of an opportunistic alliance between anti-tech forces and some hi-tech splinter group.

I hope Hugo's dark vision never comes to pass — but I can't deny the possibility that some sort of violent conflict could one day erupt out of the dichotomy between

- growth into the Cosmos
- stricter preservation of the traditional boundaries constituting humanity

This dilemma will almost surely be obsoleted in various ways, as time moves on — but the path to obsoleting it might not entirely be pretty. Certainly, there are heavy shades of the "growth versus preservation" issue in various violent conflicts on the present geopolitical scene.

But difficult conflicts have been there all through the evolution of life, complexity and mind on Earth. The early mammals surely struggled mightily day by day, even though in the big picture they were obsoleting the old order and laying the basis for the creation of all manner of exciting new life and intelligence.

By focusing on the glorious positive possibilities of the future, one isn't denying the need to pay attention to the sometimes harsh realities of the present. But one *is* denying the wisdom of becoming overwhelmed and oppressed by present realities. Getting overwhelmed by present realities leads to problems like focusing on building weapons or crafting advertisements, rather than creating longevity therapies, building beneficial AI systems, or educating children.

We live in a region of reality in which the great Cosmist truths are hard to perceive sometimes — but that doesn't diminish their relevance and importance. The challenge is to keep the deeper picture always in mind while also fully engaging with the (sometimes frustrating, sometimes terrifying, sometimes amazingly wonderful) realities of the corner of the Cosmos that presently confines and defines us....

Ultimately, the good and bad aspects of the human reality we live in are different aspects of the same thing — the same human nature, which is a particular manifestation of universal nature (a particular way of separating the Cosmos from itself!).

So it's never really "Us versus Them", it's always "**We versus Us.**"

Though it seems apropos to quote (the great gypsy punk band) Gogol Bordello here: "*We know there is no us and them; but them they do not think the same.*"

Coming to terms with the various conflicts between humans and other humans, is really a part of the process of human nature coming to terms with *itself*.

Which is part of the process of human nature growing beyond itself – of humanity overcoming its separation from the remainder of the Cosmos –

# The Power of Positive Intuition

Returning to the meme of Cosmism as post-religion, it's worth noting that one of the big strengths of traditional religions is their ability to harness the "Power of Positive Thinking" (a phrase that Americans of a certain age will recognize as the title of an influential self-help book, which I found a quick, amusing and moderately inspiring read during my early teenage years).

It's remarkable how much impact can be obtained, among us humans, by Believing You Can Do It. The power of faith isn't unlimited, but it's tremendous. The placebo effect illustrates this, as do the stirring anecdotes in the self-help section in the bookstore, and the practices of martial arts and athletic coaching. As Mohammad Ali liked to say, "How could I fail, with Allah on my side?"

And the power of positive thinking is well demonstrated from a scientific as well as a subjectivist perspective. Even setting aside the (rather strong) evidence for psi, there is ample evidence (in peer-reviewed studies) for the impact of the human mind's expectations upon the human mind and body's performance. Some of these studies highlight the subtlety of the impact of minimal positive cues: for instance, people will do better at solving math problems if they first read a story about a mathematician who shared their birthday, as opposed to a mathematician who did not.

At the 2007 conference of the World Transhumanist Association, I gave a lecture focusing on the transhumanist relevance of the power of positive thinking. It was called "A Positive Singularity in Ten Years — If We Really, Really Try" – and the video of that talk has been my most popular online video lecture.

I truly believe the title of that talk: if a significant portion of the human race passionately wanted to create a positive Singularity within the next 10 years, it would very probably get done. It was true in 2007 and it's true as I type these words in 2010. How fast might our progress toward a positive Singularity have been between 2007 and 2010, if a large number of people had been pushing very hard toward this goal?

Why did the Manhattan Project result so rapidly in the creation of effective atomic weapons, in spite of the difficult science and engineering problems involved? Because the scientists who gathered in the desert to work toward this goal, were *very very very* motivated to succeed, due to rational fear of their truly horrible enemies. And fortunately, the scientists working for Hitler's rival atomic bomb effort, were not motivated to a similar degree.

While Cosmism lacks the superstitions of conventional religions, it shares with many of them an emphasis on the power of positive thinking and feeling. When you truly want something, and believe it's possible, quite often you find ways of making it happen. This is what the word "will" means in the Ten Cosmist Convictions cited near the start of this Manifesto.

## Naturally Driving the Singularity

It may seem fanciful to consider that we can "will a positive Singularity into existence." But if one replaces "free will" with "natural autonomy" as discussed above, the case becomes clearer.

It seems eminently believable that we can *naturally drive* a positive Singularity into being.

The criteria for an action possessing natural autonomy have to do with modeling oneself as an agent of the action, and with the specific outcome of the action depending sensitively on one's own internals. These criteria are plausibly fulfilled in this case.

We can and should model ourselves as agents

participating in bringing the Singularity about –   we can tightly integrate our relationship with the Singularity into our self-models.

And it seems quite feasible that the nature of the Singularity does depend fairly sensitively on what we do at this juncture in our history.

Our attitude during the next decades may have a dramatic impact on what comes afterwards for our descendants –   and may determine such matters as

- whether we have any descendants at all

- whether our descendants include any humans

- whether we have meaningful cognitive continuity with our nonhuman descendants.

Much is at stake; and the odds of success are almost surely maximized if we step into this beckoning unknown with minds that are both clear and positive.

# Neutral Reason and Positive Intuition

Great athletes often combine reason and faith in the following way. They reason very carefully and rationally about what approach to take. But then when executing their approach, they adopt an attitude of deep belief and faith that they *will* succeed.

First, they choose.

Then, within the context of their prior choice, they flow.

This teaches a valuable lesson: the "power of positive thinking" is an inspiring but crude notion, and positive thinking works better for some cognitive processes than others. This lesson resonates closely with Cosmist ideas.

We must reason and think and analyze about ourselves and each other and our future. It is, after all, our analytical and creative cognitive powers that have created the science and technology to bring us where we are today, on the verge of transcending our legacy humanity and stepping into new domains. We must carefully consider risks and downsides, and the possibility of critical errors or violent opposition.

But reason can't do everything — we must also rely on our emotion-driven intuition. And this is one place where the power of positive thinking comes in.

In logical, rational reasoning one wants to be cool-headed; one wants to estimate probabilities without letting one's hopes, dreams and expectations muck up the calculations. People are generally not nearly as good at this as they could be given the power of their neural circuitry, and it's worth spending effort to tune one's "neural reasoning engine" better. Study of the psychology literature on "heuristics and biases" can be helpful in this regard.

But logical reasoning is not how the human mind conceives the new ideas it reasons about – and there are (strong though uncertain) reasons to believe that no mind will ever conceive radically novel, creative ideas via rationality alone. Logical reasoning is incremental by nature, whereas creative conception is holistic. Even today's crude AI systems tend to use non-logical mechanisms like neural nets or evolutionary algorithms to create new ideas.

Intuition, which proceeds holistically and conceives (sometimes surprising) new ideas, is not meant to be impartial and unbiased in the same sense that logic is. It is largely driven and directed by attitude. Like an athlete's execution, intuition is a place where the power of positive thinking shines.

And so it is imperative for us to encourage our individual and extended minds to fully embrace the positive possibilities that Cosmism indicates. We need to reason coldly and accurately about the world we are helping create – and we also need to intuit about it with positive creativity.

Believing doesn't necessarily make it happen — but appropriately mixing passionate intuitive belief with careful rational consideration may dramatically increase the odds.

## The Secrets of Individual Greatness

Some people achieve massively more than others in life — and in recent years a fair amount of empirical study has gone into understanding why.

The following seem to be the key ingredients:

- **Work very hard**. Practice a lot. It typically takes 10,000 hours of practice to become really good at something, whether it's a sport, an art form, or an intellectual or social activity.

- **Practice intelligently and creatively**: constantly challenge yourself in novel ways. Push yourself to do things a bit beyond your limits; and constantly test yourself against new sorts of obstacles. Pay attention to your performance and understand your weaknesses and strengths, and modify your practice according to your understanding.

- **Model yourself accurately** insofar as you can, with an explicit goal of improving yourself.

- If you want to be creative as well as masterful, then **practice creativity explicitly**. Think differently — then differently from that, etc. Take time each day to practice coming up with wacky ideas. Invent unashamedly, in the manner of brainstorming, without worrying about the quality of your products.

- **Adapt to your own strengths and weaknesses —** unless you have a truly abysmal lack of talent for a certain area, there are probably ways to become masterful at it that amplify your particular strengths and avoid your weaknesses.

- **Mix cold hard reason with faith-full, positive-thinking-imbued execution and intuition**

This recipe may sound overly simplistic — but we all know it's not so easy to achieve such things within the constraints of the human motivational system. Persistence itself is difficult enough to come by; but even more so, persistence coupled with ongoing cleverness and creativity and positivity and focused attention over thousands of hours of practice.

Matthew Syed's book *Bounce* contains an interesting review of several of the above points; and for the viability of teaching creativity, see Edward de Bono's classic works on brainstorming. Both of these authors emphasize the point that in Western culture we place far too much emphasis on innate talent, and far too little on the sorts of ingredients that I've mentioned above.

The bottom line is: Unless you're old and on the verge of death, or have some extraordinary disability, the odds are high that you can achieve mastery and greatness in whatever domain you choose — if you really want it enough to pursue it with true persistence, passion, attention and adaptability.

# Childhood Patterns of Great Achievers

None of the above " secrets" tells you how to *find the motivation* to push yourself in the ways that are likely to lead to dramatic success! But there is also some data on where great achievers tend to get their inspirations. I discussed above that sometimes madness can inspire great passion and creativity; but fortunately there are other routes as well!

Decades ago, my father and his parents wrote two books on the childhoods of famous people (*Cradles of Eminence*; and *Three Hundred Eminent Personalities*), and among their conclusions were that

- famous scientists tended to have scientist-mentors in their youth, plus long stretches of time for pondering and self-study
- famous artists and writers tended to have troubled youths wracked with emotional conflicts

So there were definite childhood patterns correlated with having motivational structures leading to the "obsessively intelligent" practice-patterns corresponding to dramatic success. Although, it must be stressed, these are merely strong statistical trends — these childhood patterns are nowhere near *guarantees* of someone being motivated to act in ways that will result in great achievements!

Another perspective to take is that of memetics — the evolutionary spread of memes or "idea-complexes." Mathematics is a meme, literature is a meme — and if someone has the appropriate preconditions, this meme may take hold in their mind. Sometimes the meme is propagated explicitly, as when a parent begins coaching their child in sports at age 4, or the case of Norbert Wiener, the father of cybernetics, whose father intensely home-schooled him in mathematics and science for his whole childhood. Other time it passes itself along in a less predictable manner. The meme is not just a complex of ideas — it's a complex of motivations as well, which urges the meme-host to devote increasing amounts of his time and life to the complex of ideas!

Ultimately, whether to pursue individual greatness or not is an individual decision. We've seen above that the notion of " decision" is something our culture typically convolutes – but this doesn't have to confuse the practical matter. In the same sense that any of us decides to get up out of bed in the morning, we can decide to orient our lives in a way highly probable to lead to overwhelming success.

## The Future of Great Achievement

One implication of this understanding of mastery is that, once the human motivational system is upgradable, we will all be able to become masters and geniuses at almost everything we want to. It will merely be a matter of tweaking ourselves to WANT to practice long and hard enough and in the right way.

Achieving "greatness", within the constraints of our physical infrastructure (to take an objectivist view) or the constraints of our integrity as an individual mind-system (to take a more subjectivist view), will be even more plainly a matter of "choice" than it is today. The costs of pursuing the path to dramatic success will be far less.

# Transhumanist Short-Cuts to Mastery?

And the plot gets thicker —

As technology advances, there may also arise the possibility to achieve mastery without all the work, by simply plugging into your brain a knowledge of expert ping-pong, or differential equations, or massage.

This is exciting, but also somewhat subtle.

When gaining mastery of an area through years of effort, you integrates that area with your whole self — you shifts your whole way of thinking and understanding accordingly.

Learning mathematics doesn't just give you certain skills at manipulating symbols — it also morphs your whole mind into a mode more focused on precise definition, cognition and argumentation.

Learning physics doesn't just give you skills at modeling physical situations with equations — it teaches you a quantitative-modeling mindset that pervades all your activities, which is one reason why physicists are so popular as "quants" on Wall Street.

Learning massage doesn't just give you physical skills,

it involves you in particular sorts of interactions with people over a long period of time, which affects your underlying attitude toward, and way of relating with, others.

Learning soccer doesn't just give you more ability at that particular game, it teaches you a lot about working with others, which is why team sports can be particularly good preparation for some kinds of business

What isn't clear is how many of these "ancillary benefits" of learning would come along from importing knowledge into your mind in a direct way using future technologies.

If you imported external knowledge in an inflexible way, the experience would be more like having an external piece of software plugged into your brain — say, a super-powerful version of Mathematica (for the math case), or a muscle control program that takes over your body and lets one give expert massages.

But if you imported external knowledge in a sufficiently flexible way, then presumably the adaptation of the rest of your brain to the new knowledge would occur gradually over time, as you used the new knowledge. Initially you might have the technical capabilities of an expert mathematician or masseuse, without the changes to attitude and self-model that are normally correspondent to these capabilities; but over time these other changes would likely come.

A sufficiently advanced technology could potentially modify your mind into something like "what your mind would be like if it had spent 10 years learning advanced mathematics," etc. In this way the integration of the mathematical knowledge with your self would be achieved all at once. But this seems to fail the continuity test — it seems more like killing yourself and replacing yourself with a superior, more knowledgeable being.

If one wishes to maintain a strong continuity of experience, it seems it will be necessary to take some time to let newly incorporated knowledge modules get fully integrated into oneself via practice. This will not require the same level of persistence, hard work and passion that achieving mastery requires now — but it will surely be a fascinating and intense experience all its own.

# Religion and Post-Religion

Religion has played a large role in the evolution of human culture — for good and for ill, and in such complex ways that dissociating the good from the ill is barely a meaningful pursuit.

But it seems clear that religion's era is fading. Science and technology do not disprove traditional religions, but they are conceptually disharmonious with it, to an extent that religion becomes less and less of a factor in the world, decade by decade.

In some parts of our world — even some parts of advanced nations like the US — religion still holds powerful sway. Even today it is very difficult for an open atheist to get elected to the US Congress, though the presence of many "undercover" atheists is clear, and there is at least one open atheist there now (Representative Pete Stark, from California). But still, if one charts the trends decade by decade, one finds that as technology spreads, religion's influence progressively decreases, except in some small local regions.

In many ways the dwindling of religion is a good thing — there have been too many religious wars; and too much suppression of joy, growth, choice, knowledge and understanding on religious grounds.

Yet religion has also brought much hope, joy and growth to many — and socially activist churches have given additional positive life-choices to many disadvantaged people.

Religion gives people a feeling of connecting with something beyond the self, in a way that science and technology don't do — it delivers a feeling of "massively extended self", of connectedness with a universal mind.

And it provides a way of focusing the "power of positive thinking", which science does not currently offer (even though science now acknowledges this power in some respects). Plenty of great things have been achieved during human performances fueled and optimized by faith that one or another God is on the performer's side.

Cosmism is not a religion. But it has the potential to deliver some of the benefits of religion in a manner more consilient with science.

# Religion, Inanity and Insanity

I recently saw the documentary film *Religulous* — a comedy with a serious message, focusing on some of the things that religious people believe, which seem absurd from a scientific perspective. The film gave overly short shrift to the social good and spiritual and psychological growth that religion can bring. But it did a great job of highlighting the fact that most religious beliefs would seem insane if held by an individual rather than a large group. Evolutionary biologist Richard Dawkins has tirelessly promoted this same view, for instance in his book *The God Delusion*, bringing atheism into the mainstream of American culture (and also associating it with a rather shrill and mocking attitude, which has itself been parodied!)

But though their attitudes may rankle sometimes, it's hard to dispute the basic point of these atheist advocates.

After all: Is Jesus hearing the voice of God really all that different from flamoot hearing an implant in his head?

Is there a big difference between the conspiratorial delusions of the paranoid schizophrenic, and the Muslim notion that "The smallest reward for the people of Paradise is an abode where there are 80,000 servants and 72 wives, over which stands a dome decorated with pearls, aquamarine, and ruby, as wide as the distance from Al-Jabiyyah to Sana'a"?

Why is the Scientology belief of a posse of aliens controlling our minds, any more insane than the Christian folk belief that the individual mind is pulled one way by God up in Heaven and the other way by the Devil down in Hell? Why are any of these saner than Dawkins' faux religious belief in a Flying Spaghetti Monster?

What about the Church of John Coltrane, which holds that Coltrane's sax playing was divinely inspired, and structures church services around is music? (I've never been to a service, but I imagine it would be more enjoyable for me than most, as I much prefer bebop jazz to gospel music or Gregorian chant!)

Each of these religious beliefs (except the Flying Spaghetti Monster, so far as I know) is considered sane by part of our contemporary society, and insane by another part.

This is a disharmony in our cultural mind that is likely to be remedied during the next century, as traditional religious beliefs fade and the scientific world-view rises.

## Avoiding the Dilemma

I recall asking a friend who is both an innovative physical scientist and an Orthodox Jew, how he could maintain both belief systems simultaneously. How could he believe he would go to God after he dies, and also study the dynamics of cognition as related to brain lesions, and the foundations of physics? It wasn't that I found the belief systems logically inconsistent — I knew he could make up stories rendering them compatible. It was that I found the combination conceptually and emotionally bizarre.

What he said was that he kept the two different aspects of his life in different modules of his mind. His scientific world-view was something he analyzed and thought about. His religious beliefs were something he avoided analyzing on principle — they were a matter of faith, not thought.

It wasn't that he took his religious faith as a kind of "simplicity measure" guiding his scientific inference. The intuitive sense of simplicity underlying his scientific work was about the same as that of every other scientist. Rather, he compartmentalized, to an extent that astounded me.

This seems to be a good metaphor for what American

society does now — with its obsession with science and technology, paid for with money stamped "In God We Trust."

We are "avoiding the dilemma" — with an internal disharmony that is characteristic of some mental illnesses, in which different parts of an individual's mind don't communicate with each other hardly at all (dissociative identity disorder and post-traumatic stress syndrome come to mind).

This kind of separation is almost never an optimal use of resources — it squanders the potential for synergy that's implicit in the rich interaction of different parts of a system. So if one is on a path of seeking to maximize growth, joy and choice, one is not likely to persist with such modularization. And indeed the odd conflation of science and technology with religious superstition seems to be disappearing, gradually, as the scientific replaces the religious world-view.

# Cosmism versus Unitarian Universalism

Another approach to conciliating religion and modern science-powered culture would be to modify religion so as to make it more agreeable with other modern ideas.

My father often attends the services of the Unitarian Church, and when he read an earlier version of this Manifesto, he pointed out some strong similarities between Cosmism and modern humanistic religions like Unitarianism — though also some differences.

While Christian in origin, to the outsider Unitarianism seems sufficiently abstracted (or some would say "watered down") that it almost could have arisen as a variant of Judaism, Islam, Buddhism, Taoism or some other religion or spiritual discipline instead. While Unitarian literature does refer to God, individual Unitarians need not believe in any kind of God — Pete Stark, the first openly atheist member of the US Congress, is a member of the Unitarian Church, a fact that may make his atheism more acceptable to his constituents.

Some basic principles of the Unitarian Church are as follows:

- The inherent worth and dignity of every person;
- Justice, equity and compassion in human relations;
- Acceptance of one another and encouragement to growth;
- A free and responsible search for truth and meaning;
- The right of conscience and the use of the democratic process within our organizations and in society at large;
- The goal of world community with peace, liberty, and justice for all;
- Respect for the interdependent web of all existence of which we are a part.

There is nothing inconsistent with Cosmism here. But, aside from the emphasis on narrower political and ethical principles than Cosmism chooses to emphasize, there is a distinct lack of attention paid to the possibility of growing beyond the reality of legacy humanity.

Cosmism, in the version I've presented here, could be closely approximated as a combination of Unitarianism and transhumanism. But the combination of those two leads to all sorts of synergetic effects that aren't obvious from either of the ingredients — and these synergetic effects are what I've tried to emphasize most heavily in this Manifesto.

# Obsoleting the Dilemma: Cosmism as a Post-Religion

Perhaps some form of Cosmism could serve some of the roles traditionally played by religion?

Perhaps Cosmism could serve as a way of maintaining some of the positive features that religions have brought humanity, without the aspects that appear "insane" from the scientific world-view.

This would be another way to obsolete the dilemma between religion and science – and could potentially have dramatically positive effects, if done right.

Certainly, it seems feasible for a group of people to band together to amplify their self-awareness and their mutual feeling of connectedness to the Cosmos — to boost their mutual joy, growth and choice — without the superstitions and trappings associated with conventional religions.

Similar ideas have been bouncing around the Internet for the last few years, including discussions within tiny futurist splinter groups like the *Order of Cosmic Engineers* and the *Turing Church* (note for non-computer-geeks: Alan Turing and Alonzo Church were two of the founders of computer science; and the idea that every describable procedure is equivalent to a computer program is called the "Church-Turing Thesis").

It may be too early for this idea to garner mass appeal, as human technology has not yet reached a point where its Cosmist implications are obvious to the masses. But — it's hard to tell — the time may be ripe for it to appeal to a "fringe" of future-savvy thought-leaders.

What might a Cosmist post-religious service look like? A combination of "Burning Man festival with more hi-tech and less drugs and debauchery" with "a futurist conference with fewer speeches and more multimedia"? This would be fun, and may end up being part of the story, but such events would seem to lack the sense of purpose that traditional religious gatherings have.

Perhaps it will be more broadly appealing to link Cosmism with the quest for individual greatness. Indeed this comes close to the perspective taken by Extropy, the first really active modern transhumanist organization, which was founded by Max More and Tom Bell in the

1980s and closed its doors in 2006 (declaring success in its mission of spreading a certain set of futurist memes). Extropy was about seeking the transhuman future, and also perfecting oneself in the present.

One or more Cosmist organizations focused on bringing individuals together to collectively help each other work toward individual greatness, building toward Cosmist goals –  this would seem to have potential.

## Coming Together as Cosmists

The possibilities of Cosmism as a post-religion bring us back to my words at the end of the Introduction to this Manifesto:

*My hope is that you'll find the practical philosophy I articulate here not only interesting but also compelling. Cosmism isn't just about cool ideas that are fun to think, talk and write about. It's about actively trying to understand more, actively trying to grow and improve and collectively create a better Cosmos, and all that good stuff...*

*As will become clear to you if you read the rest of this Manifesto, one aspect of Cosmism is, that, roughly speaking: the more sentient beings adopt Cosmist values, the better will Cosmist values be served.*

*Of course, I don't expect anyone to fully agree with everything I say here — I myself, in a decade or a year or maybe even a month, may not agree with all of it!*

*However, if you agree with a substantial percentage of Cosmism as I articulate it here — and more importantly, if you agree with the **spirit** in which these thoughts are* *offered — then you are a Cosmist in the sense in which I mean the term.*

Perhaps many of the ideas in this book resonate with you. Perhaps we are both, in similar senses, Cosmists. What are we going to do about it?

We can each work, individually, toward Cosmist goals. But are there ways we could join forces to proceed more successfully together?

One of the things that religions have done well is building social unity. Folks with the same religious world-views and value-systems tend to stick together, helping each other out practically and advocating for the same causes, etc.

Wouldn't it be nice if Cosmists did the same?

It's a fact that, in human history, *rigid, inflexible bodies of ideas* tend to be particularly good at at attaching themselves to *well-organized, well-coordinated bodies of people*.

For instance, in the early days of Christianity, there were loads of interesting, egalitarian, semi-anarchic Gnostic sects ... but none of them achieved the ultimate staying power and influence of Catholicism. Why? Because Catholicism had a belief-system that coordinated well with an authoritarian social structure ... which led to more effective recruitment into and propagation of the religion.

Cosmism is open-ended and flexible by nature, hence would not lend itself well to a hierarchical, authoritarian organization. There will never be a Cosmist Church of any size or influence.

But, perhaps, some sort of *confederation of Cosmists* could emerge....

It could emerge within some existing futurist organization, such as Humanity+, as a sort of focus group of H+ adherents passionate about Cosmism in particular. Hypothetically it could emerge within some sympathetic religious organization such as the Unitarian Church. Or it could emerge as a novel organization – the Order of Cosmic Engineers and Turing Church being closely related examples. My own feeling is that the seeding of a Cosmist confederation within a religion or something with a "Church" label is not likely to be the optimal route, though it's possible this just reflects my personal biases, and I'd be happy to be proved wrong.

Top universities and athletic academies show that gathering together like-minded individuals in pursuit of individual and collaborative excellence, can be an effective strategy. What if the same sort of energy were spent building a tightly interacting community of individuals with mastery at working toward Cosmist goals, as is now put into building communities of great electronics engineers or baseball players?

Whatever its organizational formalities, such a confederation would need to be very comfortable with the possibility of its own impending obsolescence.... For one thing, as technology and science advance, and we expand and advance our minds accordingly, Cosmism in some form may become simply common sense ... so that the confederation of Cosmists would then consist of all sentient beings. Or on the other hand, once we expand our scope of understanding a bit, Cosmism may come to look as quaintly obsolete as ancestor-worship does to us now.

But the task now, for those of us who do broadly accept Cosmist ideas, is to work to make sure a positive future comes about: one in which increasingly intelligent and able minds bring about escalating levels of joy, growth and choice for a rich variety of sentiences. It is worth thinking hard about what we can do toward this end: in terms of optimizing our own paths to individual greatness, and in terms of banding together informally or in more organized manners. The stakes are certainly high –  as high as our human minds are able to comprehend, and possibly far higher.

Thanks for listening ...

*Onward and upward!!!*